Joann Karges

Nabokov's Lepidoptera:
Genres and Genera

Ardis, Ann Arbor

Cover: *Parnassius mnemosyne* by Sheri Williamson.

Joann Karges, *Nabokov's Lepidoptera*
Copyright © 1985 by Ardis Publishers
All rights reserved under International and Pan-American Copyright Conventions.
Printed in the United States of America

Ardis Publishers
2901 Heatherway
Ann Arbor, Michigan 48104

Library of Congress Cataloging in Publication Data

Karges, Joann.
    Nabokov's Lepidoptera.

    Bibliography: p.
    1. Nabokov, Vladimir Vladimirovich, 1899-1977—
Knowledge—Zoology.   2. Lepidoptera in literature.
I. Title.
PS3527.A15Z74   1985   813'.54   84-24601
ISBN 0-88233-910-9

To my son, John

Contents

Preface

When I first began to consider the Lepidoptera that Nabokov uses in his literary works, including the interviews which I consider "literary" by his carefully pre-planned statements, I was excited by the sheer diversity of his Lepidoptera. My own activities in Lepidoptera (far, so very far inferior to his) prompted me to explore further the subtleties of his usage of them.

As every in-depth reader of Nabokov soon realizes, all allusions, phrases, names, and words are deliberately and consciously selected and often bear double meanings, even layers of meanings, some of which are apparent only by subsequent references in the text. The reader's participation is invited to a puzzle that lies beneath the narrative. With this premise for the reading, his Lepidoptera offer some surprising interpretations. It is not, however, only the butterfly or moth as particular insect that is involved in his use of Lepidoptera. Repeatedly, especially in the interviews, he has given us strong clues to his chief Lepidoptera interests, which scholars have been inclined to relate to his scientific articles alone. This interest is taxonomy, or more precisely, systematics of the order, abundantly revealed but often disguised, in his novels.

From the beginning of Nabokov scholarship a number of otherwise knowledgeable critics have attempted to capture and pin his Lepidoptera, and some have made very erroneous statements about them—from the eclosion of the insect to their flight behavior and their food preferences. These I do not intend to refute or even correct. I hope, rather, to enlarge the understanding of Nabokovians (we are all that who admire his literary works and his scientific contributions to Lepidoptera) and to elucidate this extraordinary fusion of the sciences and arts of Lepidoptera and systematics and literary talent in these incomparable works.

9

One's first acknowledgment, I believe, should go to one's author. I am grateful not only for the experience of these masterpieces and the remarkable mind of their author, but also for his having introduced me through his writings to the marvels of Pushkin and other Russian poets and to the excitement and aesthetic appreciation of zoologic systematics. My understanding of each can but grow.

Additional appreciation is extended to all of the Nabokovian scholars whose works are cited in my bibliography of secondary sources and who have, each in his own way, contributed to our understanding of this multifaceted genius. I am equally appreciative of the continuing research and published works of many lepidopterists, but especially I am indebted to Dr. Robert M. Pyle for invaluable assistance in correcting the manuscript and for advice and encouragement.

I thank Mrs. Vladimir Nabokov for graciously permitting me to quote from the works of Vladimir Nabokov.

Acknowledgments

To Mrs. Vladimir Nabokov for permission to quote from:
The Eye (Phaedra Press, c1965); *Laughter in the Dark* (New Directions, c1938); *Nikolai Gogol* (New Directions, c1944); *Pnin* (Doubleday, c1959); *Poems* (Doubleday, c1959); *The Real Life of Sebastian Knight* (New Directions, c1941); *Speak, Memory* (Putnams, c1966); "A Few Notes on Crimean Lepidoptera" (1920); "Notes on the Lepidoptera of the Pyrenées Orientales and the Ariège" (1931); "Some New or Little Known Nearctic *Neonympha*" (1942); "Notes on Neotropical Plebejinae" (1945).

To Mrs. Vladimir Nabokov and McGraw-Hill Book Company for permission to quote from:
Ada (c1969); *The Annotated Lolita* (c1970); *Bend Sinister* (c1947, 1974); *Despair* (c1965); *Details of a Sunset, and Other Stories* (c1976); *Glory* (c1971); *King, Queen, Knave* (c1968); *Look at the Harlequins!* (c1974); *Mary* (c1970); *Poems and Problems* (c1970); *A Russian Beauty, and Other Stories* (c1973); *Strong Opinions* (c1973); *Transparent Things* (c1972); *A Tyrant Destroyed, and Other Stories* (c1975).

To Mrs. Vladimir Nabokov and the Putnam Publishing Group for permission to quote from:
The Defense (Capricorn Books, c1964); *The Gift* (Capricorn Books, c1970); *Invitation to a Beheading* (Putnams, c1959); *Lolita* (Putnams, c1955); *Pale Fire* (Putnams, c1962).

To Mrs. Vladimir Nabokov and Harper & Row for permission to quote from *The Nabokov-Wilson Letters, 1940-1971,* edited by Simon Karlinsky, c1979.

To Princeton University Press, Bollingen Series for permission to quote from *Eugene Onegin,* by Alexander Pushkin, translated with a commentary by Vladimir Nabokov. (Bollingen series LXXII, c1964).

To Columbia University Press for permission to quote the passage from *Systematics and the Origin of Species,* by Ernst Mayr, c1942, on p. 59.

To Alfred A. Knopf for permission to quote the passage from *Butterflies* by Thomas Emmel, c 1975, on p. 50.

1

Nabokov, Lepidopterist

Throughout his interviews and his letters, Nabokov emphasizes his interest in the scientific study and classification of Lepidoptera. The reader should be wary of such statements as Nabokov could be deliberately deceptive and because a non-lepidopteran reading of his literary works belies such interpretation. Far exceeding mere signatures of authorial identity, his Lepidoptera are at times poetic imagery and symbolic, but even more importantly, they are theme, pattern, and texture. He uses butterflies and moths in conjunction with other patterns—time and memory, microscopic vision, the camera lucida, mimicry or natural deception, and the spiral. As he requires much of his readers in being knowledgeable of languages and several bodies of literature and in exerting their imaginations to understand puns, artistic and literary allusions, and chess problems, so he intimates that they, too, should understand Lepidoptera and to some degree zoologic systematics, because these he also uses as artifice in his writings.

In an interview with Alvin Toffler, Nabokov outlined the chief appeals he found in his Lepidoptera activities: the hope of capturing a specimen heretofore unknown to science; the capture of a rare or local species known to the capturer only from plates or descriptions in books; studying the life histories of little-known insects and placing them in correct classification according to their physical structure; and the sport itself, the goal being the specimen in the little envelope for later study.[1] Each of these points is illustrated in his poems, stories, and novels, some many times, as well as in *Speak, Memory*.[2] For example Godunov-Cherdyntsev (*The Gift*)[3] was author of a new species; Kanner (*Look at the Harlequins!*)[4] explains the distinguishing features of his rare find, the Ergane butterfly; Ada assiduously studies life histories of her Lepidoptera;[5] and in *King, Queen, Knave*[6] Dreyer calls butterfly-catching "good sport."

It is in *Speak, Memory,* however, that Nabokov expresses a deeper meaning in his Lepidoptera preoccupation, related intrinsically to his art: "a form of magic," a "game of intricate enchantment and deception" (p. 125), and ecstasy and "something else" (p. 139). The study of structure for establishing order in organic diversity through systematic classification is a scientific value and adds validity to collecting activity. The progression of Lepidoptera through instars to mature insect, while not having exact correlative in the Hegelian triad, is a "becoming" process that synthesizes in beauty, with supreme examples in butterflies and many moths but in some ways even in the plainest of, say, cutworm moths (Noctuidae). The concept of nature's mimicry exceeding its purpose, that is, the creation of beauty in species which surpasses the function of deceiving the predator has a parallel in the beauty of the species surpassing the ultimate biological purpose, attraction of a mate for procreation and perpetuation of the species. Therefore, nature mimics art in having value exceeding express purpose. The butterfly becomes a symbol of this transcendence and also a symbol for artistic endeavor as transcendence over the solipsism that this interpretation of nature ultimately leads to.

Speak, Memory, deceptive autobiography that it may be,[7] nevertheless reveals the simultaneous development of author and lepidopterist, the concentric spiralling of interests into his scientific and literary achievements.

Absorbing the *passio et morbo aureliana* from his father, Nabokov had begun observing and collecting Lepidoptera from the age of seven or so. While recuperating from an illness at age ten, his mother presented him with a magnificent library of the best of Europe's lepidopteran writings, the first of the great illustrated works of Seitz, and those of Hoffman, Boisduval, and others (several of whom were to receive collaboration from Godunov in *The Gift*). These were to stimulate him to serious study and an interest in finding a new, as yet undescribed species. In fact, at least twice as a youth he tried unsuccessfully to be a "first describer," a goal that he was finally to realize only many years later. Meanwhile, he learned the joys of the microscope, "the precise and silent beauty," "the point arrived at by diminishing large things and enlarging small ones that is intrinsically artistic" (pp. 166-67), a concept that is evidenced in his literary viewpoint. His lessons in sketching and drawing were to stand him in good stead when he was later engaged as a lepidopterist and was writing and illustrating his scientific works. In the summer, his family left St. Petersburg for the country estate at Vyra, where he would pursue his interests with more intensity; there were also trips abroad, to Biarritz and elsewhere, which offered him opportunity for a broader acquaintance with European Lepidoptera. The degree of his boyhood activity is illustrated in his recounting the incident of Mademoiselle lowering her great bulk onto a fine series of Large Whites; in this his deepest distress was the destruction of the gynandromorph that he had collected and which would have been of particular pride to any lepidopterist (pp. 127-28).

The revolution was of course a major disruption in their lives. If it had not been for the revolution, Nabokov said, he might never have written novels but instead would have become a lepidopterist.[8] It is interesting to speculate that in that case, he, not Godunov-Cherdyntsev, would have followed Nikolai Kuznetsov[9] in contributing such works as *Lepidoptera asiatica* (in eight volumes) or *Butterflies and Moths of the Russian Empire* (in six volumes) (*The Gift,* p. 114).

Escaping to the Crimea (regretting that it was the end of the collecting season), Nabokov became acquainted with still more species that were new to him. His experiences there led to his first lepidopteran publication in 1920,[10] while he was a student at Cambridge. Little then would he have guessed that the Red Admiral butterflies "sailing and fluttering among oak tress" of that article were to become an important image in a novel forty-two years later. His experiences in the Crimea were also to be reflected in *The Gift,* in which Fyodor boxed southern butterflies, knowing that the real rarities were "among the rocks of Ai-Petrie and on the grassy plateau of the Yayla" and where Fyodor's father showed him the satyrid that Kuznetsov had recently described (p. 141).

From 1919 emigration became a pattern, an "antithesis" in the Nabokov family's life. Following schooling in England, he was a resident of the emigre[1] colony of Berlin, and with the writing and publishing of poems, stories, plays, and novels, he experienced a ten-year hiatus in his "butterfly hunting." In 1929, however, a trip to the eastern Pyrenées and the Ariège led to the publication of his second lepidopteran article[11] (and the writing of *The Defense*). Opening with the telling of a dream (and absolutely no reference to Freud) and including a typical Nabokovian literary sentence,[12] the many descriptions of Lepidoptera that would be echoed in butterfly references in his fiction reveal the writer-lepidopterist also as painter with an exquisite sense of color and attention to exact detail. There is also in the article the amusing aspect of the "man with a net" that he used in humorous elaboration in *Speak, Memory* and parodied elsewhere.

From Germany to Paris for several years and then to New York was a three-part migration (beginning of the synthesis) for Nabokov, his wife, and son—from Palearctic to Nearctic fauna, from the use of the Russian language to the English language (bridged by *The Real Life of Sebastian Knight*) to professional activity as a lepidopterist and within eighteen years to fame with *Lolita*.

During the time that he was establishing his literary career in Berlin and Paris, he had little time for lepidopteran activity. Undoubtedly he made efforts to keep abreast of current research, but in his non-scientific writings there are few references to the literature after the early 1920s. He does, however, reveal a thorough knowledge of earlier developments in the field. He makes frequent, usually somewhat disparaging, mention of practices of nineteenth-century German lepidopterists, which he partially elucidates in *Speak, Memory*. With Staudinger as "high priest" (and also a chief retailer of specimens), the emphasis was on classification of Lepidoptera by sight-similarity rather than on morphological structures, which for some time had been of interest to English-speaking scientists. In addition the identification of subspecies and geographical races became a more satisfactory method of illustrating evolutionary development and ecology (pp. 123-24). He also eschewed the "chit-chat, inaccurate observations and downright mistakes" of the renowned Jean-Henri Fabre of nineteenth-century France (voiced by Fyodor as his father's opinions in *The Gift,* p. 120). His own observations and his concerns with taxonomy are evidenced in his pre-American novels, especially *The Eye*; this aspect of the science was to be his concentration as a professional lepidopterist in the United States—and the controlling theme in the novel that brought him international fame.

Two stories of the pre-1940 period are based on lepidopteran themes.

"The Aurelian"[13] is about lepidopterist Paul Pilgram, who runs an urban butterfly store that sells more trivial merchandise to school children than butterflies, and who dreams of expeditions to faraway places to pursue his collecting activities and perhaps to discover a new species. The day comes when, by selling a large butterfly and moth collection, he can plan to leave the shop and go off in search of his ideal. An unfulfilled artist in a way and a pathetic character, he would realize his dreams only in death: the "velvety black butterflies soaring over the jungle and a tiny moth in Tasmania and a Chinese skipper . . . ." The story reflects some of Nabokov's concerns with systematics but more

16

importantly with a transcendent goal in human activity and transcendence over death, both represented by butterflies.

Collecting activities are also a part of "Christmas."[14] During this season a young boy, a budding lepidopterist, has died and his father has brought the "pear-shaped" cocoon into the warmth of the room only to see the exotic moth emerge. The use of the immemorial motif of rebirth and resurrection in Lepidoptera metamorphosis is obviously contrapuntal to the Christmas story. But the *Attacus* (Atlas) moth does not represent the soul of the son transcendent over death.[15] Nabokov uses only butterflies in this way. (John Shade in *Pale Fire* rejected the Luna Moth for this purpose.) Rather, this moth with a wing span of up to 320mm emerges from its relatively small cocoon, and the large beautifully patterned and colored wings make the body seem disproportionately slender, suggesting for the purpose of the story that beauty and art are values exceeding the temporal or corporeal. An Asiatic species, the moth has been transported far from its native habitat, like the emigre characters of the story.

In his 1927 poem "V Rayu" ("In Paradise")[16] Nabokov makes reference to his continued passion to describe a new species, a "wild angel" with eyespots, which could be a satyrid, a nymphalid, or a lycaenid, except that he alludes to its blueness, which limits it, in Europe at least, to Lycaenidae. It is not surprising that he notes "fatidic" events, because the lycaenids, especially the blues, were the major concern of his Lepidoptera activities and in the conception of *Lolita*. In fact, one of his first publications[17] in the United States was the description of a new "clear silky-blue" lycaenid he had found in France and to which he gave a distinctive name, though hypothesizing that the species might be a hybrid. For this, as for his initial descriptions of *L. a. sublivens,* he had no females to examine. There is a suggestion of this species in one of Humbert's early loves; in any case, its description is a loop in the natural spiral of Nabokov's lepidopteran activities.

A pinnacle was attained in his discovery and naming of the Karner Blue from New York, which found expression in two literary genres. Pnin stumbled across them:

A score of small butterflies, all of one kind, were settled on a damp patch of sand, their wings erect and closed, showing their pale undersides with dark dots and tiny orange-rimmed peacock spots along the hindwing margins: one of Pnin's shed rubbers disturbed some of them, and, revealing the celestial hues of their upper surface, they fluttered around like blue snowflakes before settling again. (*Pnin, p. 128)[18]*

The event itself Nabokov celebrated in the poem "On Discovering a Butterfly."[19] He describes the obvious external features that distinguish this butterfly from other species in the genus: "the special tinge," the "dingy underside and chequered fringe," and the analysis of the genitalia that characterize it as a new species: the symmetrically sloping hooks like badminton paddles. "Godfather" and its first describer is he who found and named it, and here is the essence of immortality on a pinned type specimen with a small red label.

After two years in New York, where he assisted in cataloging Lepidoptera at the American Museum of Natural History, he joined the Museum of Comparative Zoology at Harvard as a Research Fellow. The seven years he spent there were his most productive in lepidopteran research and publication and also a period of high literary activity.

Of all his important lepidopteran publications, only one concerned a family other than the Lycaenidae. His article on "the delicately ornamented, quickly fading satyrids"[20] was based on specimens he collected himself in Arizona and those from collections in museums. With the exception of the European Meadowbrowns (in *Laughter in the Dark*), he did not use members of this family in his novels and yet, in the various interviews of *Strong Opinions,* he made mention of a number of species that interested him.

His research concentrated on the Lycaenidae (the gossamer-winged butterflies), Neotropical and Nearctic, with studies of evolution and distribution that included the Palearctic genera, but his main object was taxonomy. Although he analyzed and compared minutely size, colors, markings, and wing venation, there was for him one superior basis for classification, the structure of the genitalia, which "reflects natural relationships better than do other principles."[21] In his article on the Neotropical lycaenids,[22] he postulated on the origin of the subfamily in Asia, its possible path of migration to the New World and the structural evolution that took place among the genera. With the number of Argus references (automobiles, light-shade allusions, theaters, etc.) in his novels, it is interesting to note his splitting of the established genus *Hemiargus* into two new genera *Cyclargus* Nabokov and *Echinargus* Nabokov (now relumped into the original genus).

His study of butterfly genitalia has only the vaguest possible references in his literary works except as a model for a literary genre (to be discussed in chapter 6), though in a letter to Edmund Wilson he compared his study to a literary sexual reference.[23]

Nabokov's major contribution to the science was his monograph,[24] completed in 1948, on the "silver-studded Blues," Nearctic members of *Lycaeides* and the revision of the genus and species in a classification that is in large part accepted by scientists today. Through examination of hundreds of specimens, he sorted out by means of the genitalia the subspecies of the *argyrognomon* (Holarctic) and *melissa* (Nearctic) groups of the genus and explained their distribution, including sympatric occurrences throughout North America. The monograph consists of sixty-two pages of text and nine plates, a thoroughly professional work in every way. To Edmund Wilson, he indicated that his monograph reads like a *roman d'aventures,*[25] and he found that the descriptions of the species and their structures was good training "in the use of our (if I many say so) wise, precise, plastic, beautiful English language."[26]

Of this period, he wrote that he spent up to ten hours a day over the microscope and all evening writing *Bend Sinister.*[27] He was also teaching, translating, writing his book on Gogol, publishing reviews, and was considering a revision of the text of the 1931 edition of W. J. Holland's *Butterfly Book,* which he had found unreliable.[28] Enthralled with his Lepidoptera activities, he sometimes found its rewards surpassing those of literature. He wrote Wilson:

> What are the joys of literature compared to tracking an ovipositing *Callophrys sheridani* Edw. to its food plant or boxing undescribed moths from the lobby windows of J. Laughlin's very pleasant though somewhat primitive hotel![29]

Later, he said:

> My passion for lepidopterological research, in the field, in the laboratory, in the library, is even more pleasurable than the study and practice of literature, which is saying a good deal . . . . The tactile delights of precise delineation, the silent paradise of the camera lucida, and the precision of poetry in taxonomic description represent the artistic side of the thrill which accumulation of new knowledge, absolutely useless to the layman, gives its first begetter.[30]

And yet he could state that "those miniature hooks of the male genitalia are nothing in comparison to the eagle claws of literature which tear at me day and night."[31]

The pull of literature won out, and the MCZ monograph was the last of his major scientific works on Lepidoptera. He discontinued because he found it "no longer physically possible to combine scientific research with lectures, belles-lettres, and Lolita (for she was on her way—a painful birth, a difficult baby") (*Speak, Memory,* p. 65). *Lolita* was not a discontinuance of Lepidoptera activity but an extension and expression of it.

Nabokov offered his opinion of butterflies in his fiction:

> ... when I allude to butterflies in my novels, no matter how diligently I rework the stuff, it remains pale and false and does not really express what I want it to express—what, indeed, it can only express in the very special scientific terms of my entomological papers. The butterfly that lives forever on its type-labeled pin its O. D. ("original description") in a scientific journal dies a messy death in the fumes of arty gush.[32]

The seeming dichotomy of scientific description and artistic intention is thoroughly resolved in Nabokov's literary expressions. There are relatively few descriptions of specific butterflies or moths, but these are exquisite. None is "arty gush," as our very deceptive author was quite cognizant. Every Lepidoptera allusion, or trope, and each named or unnamed species is distinctly purposeful to its text.

2

The Lesser Lepidoptera

In the translations of his Russian novels into English, Nabokov increased the butterfly imagery, Jane Grayson notes.[1] Since the translations with one exception were made after he had achieved fame in his most butterfly-famous novel, *Lolita*, the reasons for the additions are obvious. By this time he was known to the literary world for his Lepidoptera activities and for his allusions, amply revealed in the afterword to *Lolita*, and in the articles about him in popular and literary magazines as well as to the relatively limited readership of his scientific works. The butterfly had become a personal emblem, and he may have added allusions as further signatures of his authorship. It is also part of his patterning.

Rather than following the strict chronology of the composition of the novels, I have found a certain natural grouping of some of them, based on the intensity of the use of Lepidoptera or on similarities and contrasts of usage. Others will be analyzed individually. The translations, either by Nabokov or with his approbation, for the purposes of this study will be cited as part of a unifed English-language canon.

Mary; Glory; King, Queen, Knave

Mary,[2] *Glory*,[3] and *King, Queen, Knave*[4] make comparatively little use of Lepidoptera as image, symbol, or allusion, and yet some of these usages are significant in themselves, and others, reflected in later works, reveal recurrent themes with identical meanings. Nearly all have some counterpart in *Speak, Memory*.[5]

In speaking of the characters of *Mary*, Nabokov uses an image of an entomomological or similar collection. The characters "in that display box were so transparent to the eye of the era that one could easily make out the labels."[6] The novel, itself, however, has few references to butterflies, all of which are similies: a bitch with "pendulous ears that had velvety ends like the fringes of a butterfly's wings" (p. 6); "a black bow looking in flight like a huge Camberwell beauty" (p. 60). The playful felicity of these two similies is unlike the third, in which there is a shade of symbolism of the transcendent spirit in a white butterfly. Letters between Mary and Ganin pass across Russia "like a cabbage white butterfly flying over the trenches" (p. 91). A parallel occurs in *Speak, Memory,* when Tamara's letters from Russia to the southern Crimea "would search for a fugitive address and weakly flap about like bewildered butterflies set loose in an alien zone, at the wrong altitude, among an unfamiliar flora" (p. 251).

In both of these the poetic comparison of love letters to butterflies takes on deeper meaning if the reader notes that the butterflies in one instance represent the souls of soldiers, and in the other, displaced persons in a foreign land. This use of Lepidoptera imagery on more than one level is typical of Nabokov's patterns.

Many of Nabokov's butterflies, particularly pale and white ones, carry the traditional ageless symbol of the *anima,* psyche, or soul.[7] Shortlived by nature, they make sudden appearances "from nowhere," as Nabokov writes, and suggest the evanescence of a spirit departed or departing from the body. For Nabokov, the pale butterfly symbolizes the soul transcendent over death and time. It is one of his most recurrent and consistently employed images of the butterfly. Moths, on the other hand, are never so used.

The moths and butterflies of *Glory* are more numerous than in *Mary* and more important in the patterning of the novel. Several of them further illustrate Nabokov's use of a single insect, or two or more insects together, for double or even multiple meaning, in addition to whatever purpose they may serve structurally in the novel.

Shortly after Martin learns of his father's death, while walking in the park below Mount Petri, he sees a "zebra-striped swallowtail gliding past," one of alternating black and white images (a black swan, a birch, black glass of the pool), and Martin realizes that human life flows in zigzags, that his own life had made a turn, not necessarily affected by the death of his father (p. 22). The black and white or negative-positive image is established; it has reference to the butterfly itself, and the butterfly also symbolizes the soul transcendent. A contrapuntal image appears also in the black butterfly "with enviable casualness" that Martin notes from his precarious balance on the mountain crags (p. 86).[8] A poetic image added for descriptive detail, the black butterfly symbolizes fate. The author of the book is also commenting on the character of Martin, who notices butterflies and moths, whether or not he discerns symbols in them.

"It is astounding how little the ordinary person notices butterflies," he writes in *Speak, Memory* (p. 129). Not all of Nabokov's characters do; most of those who notice butterflies and moths also have a well-developed artistic sense in one way or another.

Blue butterflies assume the traditional bluebird symbolism when Martin sees them fluttering up from the road; he reflects on how hard it is to capture happiness (p. 48). To him butterflies mean joy as part of a "marvellous world" of his "viatic freedom" (p. 161).

The imagery of moths, on the other hand, help to make evident the circle of Martin's life. While traveling on the train with his parents as a child, Martin had noticed on a platform a box marked "fragile" and had seen one large moth accompanied by midges circling a gas lantern (p. 21); as a young man, perhaps on the same platform, he again notices a box so marked and "one ample dark moth with hoary margins" again in the company of midges (p. 158). The box contains insects, the results of someone's collecting activities, destined for the national musuem. The novel circles; the author evidences his control through the absent entomologist, and Martin, like the insect collector, was striving for something beyond himself, something that approximates art.

There is a reflection of this allusion, too, in *Speak, Memory,* in the scene glimpsed from the train window: "like moons around Jupiter pale moths revolved about a lone lamp" (p. 146). Moths do not simply circle around light; they spiral.

Like Martin, Dreyer in *King, Queen, Knave,* also appreciates life, breathing, and beauty, and he is an artist. When Franz, Martha, and Dreyer see the couple with the butterfly net (a parody of himself and his wife, Nabokov reveals in the foreword to the book), he says: "In fact, I think to have a passion for something is the greatest happiness on earth" (p. 233). It is natural that he should notice the Red Admiral, and even though "the dark-brown ground was bruised here and there, the scarlet band had faded, the fringes were frayed," like lives disappointed and loves betrayed, he sees the butterfly as "lovely and festive" (p. 44). A momentary symbol of all that he believes in, it also predicts his betrayal in love by Franz and Martha.

The soul-departing symbol is used for Martha when she lies struggling for life in the hospital, and Franz notes a white butterfly "battling the breeze" (p. 260). There are other signs of change and transparency on the beach, but it is beyond Franz's capability to recognize either a fleeing spirit or mutability. The author's intention for him *not* to recognize images is indicated by the presence of the butterfly-hunting couple in that scene.

Despair; The Defense

There are even fewer Lepidoptera references in *Despair*[9] and *The Defense*[10] than in the three novels just discussed, and their usages are quite different.

The "small blue butterflies settling on thyme" occur when Hermann sees snow and leafless branches on trees on that June day when he, Lydia, and Ardalion are on excursion to Ardalion's woodland property, later to be the scene of crime. "The future shimmers through the past" (p. 48). These blue butterflies do not signify happiness; they indicate that neither reader nor characters should have any doubt about the season despite Hermann's confusion. The butterflies are not only feeding on thyme blossoms;[11] it is breeding season, and they have already mated. The females lay their eggs on these plants—the first step in the life cycle that leads to metamorphosis. Hermann is still in the early planning stage of his plot.

Later in the novel, Hermann tells Felix that as a child, he picked from the rosebushes caterpillars that resemble twigs. These larvae, also called span- or inch-worms or loopers, of geometrid moths, are one of the many indications of duality that is a central theme of the novel. One of nature's protective devices, it is to the human eye, and even more importantly, to the predator, a deception. A natural deception, unlike the mirrors of the novel, it involves behavior on the part of the insect. As imagery, the larva also prefigures the all-important stick that Hermann overlooked. Mimicry in insects was an aspect of nature that greatly interested Nabokov and one that he translates into art through Lepidoptera in various ways.

Luzhin's father (*The Defense*) has a curious series of sensory experiences shortly after he has lost the chess game to his son and has realized that his son is playing seriously "as in a sacred rite" (p. 66). "The fat-bodied, fluffy moth with glowing eyes" that collided with the lamp and fell on the table foreshadows what Luzhin is to become: fat, soft and ineffectual as a person, and clumsy; he is to fall in his ineptitude, his "breakdown," and literally in his eventual suicide. The breeze in the garden and the clock's chimes indicate change and inevitability, the limits of chance.

"Nonsense," he said, "stupid imagination."

Few of Nabokov's characters, major or minor, perceive and acknowledge symbols. For the most part it is left to the reader to make the associations.

Laughter in the Dark

In *Speak, Memory*, Nabokov tells us that he got even with the lepidopterist Kretschmar who preceded him in describing a moth by giving his name to a blind man in one of the novels (p. 134). The name was changed to Albinus in the English translation.[12] *Laughter in the Dark* has both moths and butterflies and expands the use of symbolism.

The use of white butterflies, plural, and a single white moth after the consummation of love in this novel suggest a single symbolic meaning. Instead, they have opposite purposes. When Albinus sees the white butterflies, he is rejoicing in the May morning, ecstatic that he has realized a dream of years' standing, a night with Margot, and the insects fluttering as though in a rustic garden suggest the erotic paradise that Nabokov uses in *Ada*. At this point, Albinus is not yet blinded by the very love that the butterflies symbolize (p. 83).

On the other hand, the lone white moth appears after Margot and Rex have had sex in the hotel's adjoining room (p. 206). The moth symbolizes not sexual passion but love deceived, especially since it falls on the white tablecloth. The deception would be scarcely visible to Albinus as white on white, predicting his physical blindness but also his inability to discern Margot's and Rex's treachery or his own blindness in love.

A clumsy moth around a rose-shaded lamp has similar meaning (p. 116). Margot, snake-like, at once temptress and predator, had been trying to catch crickets, and Albinus has recognized his jealousy. This moth too represents deception in love, actual or predictable, but Albinus sees neither himself nor the symbolism in the moth.

Neither the moths nor the "white butterflies" are identified; there is, however, a named species. Visual and poetic detail are apparent in the description of the radiator "crammed with dead bees and dragonflies and meadowbrowns" (p. 201); the significance is in the choice of the species of butterfly. To some degree all of the insects (though dead) typify the characters: Rex in the bee that seeks flowers; the dragonfly, predacious like Margot; and the rather drab common Meadowbrown Butterfly. But, in each corner of its upper forewings, this species has a single ox-like eye, sometimes called "blind-eye." It is, of course, a repetition of the ocular references to Albinus's blindness, physical and in matters of love.

Transparent Things; Look at the Harlequins!

After using Lepidoptera extensively and intensely in *The Gift* and later in *Pale Fire* and *Ada*, Nabokov reversed himself and parodied his own use of Lepidoptera in his last two novels. Both Hugh and Vadim ostensibly dislike insects, and like Humbert Humbert, become anti-lepidopterists, tantamount to anti-heroes in Nabokov's writings.

Transparent Things[13] contains the familiar white butterfly that symbolizes Armande's soul. Blotched, faded, the transparent margins of its wings of an "unpleasant crimped texture," it is especially unlovely to Hugh; furthermore, he does not attach supernatural significance to the white butterfly. Even though he had been looking for the exact spot where he and Armande had shared a memorable kiss, the possiblity of symbolism does not occur to him. In kindness, though, or as William Rowe puts it, [14] Hugh, affected by Armande's spirit, would transport the butterfly to a flower where it could feed and rest, but the butterfly also eludes him and "vigorously sailed away" (p. 90).

Other butterflies in this novel are imprisoned and mounted as decorative objects, a degenerative use, which will be discussed in Chapter 8. The important statements that Nabokov makes in *Transparent Things* are not expressed in Lepidoptera. Nor are they in *Look at the Harlequins!*

In Spain there is a butterfly known as the Arlequin, the Festoon (*Zerynthia rumina*). The harlequins of *Look at the Harlequins!*[15] are, among other things, butterflies, though Vadim emphatically insists that he knows nothing about them and would cringe at having one of the "fluffier night-flying" ones touch him, while even the prettier ones make him shiver (p. 34). Nevertheless, he *sees* butterflies, and three of his loves are associated with a butterfly sighting.

Vadim is with Iris when he sees the Pandora: "a tremendous olive-green fellow with a rosy flush somewhere beneath," an imprecise but reasonable description of the over-all olive-green cast of the underside of the wings with the rosy shading on the upper ones, but this butterfly is "speckled" on the upper side. Vadim had thought he was seeing two species; in fact, he saw only one. The Pandora has no mythological or lepidopteran relation to Iris, however. The species is used, rather, as a repeated pattern of the philosophical love poem that Vadim has read and explained to Iris. "Being in love" (*vlyublyonnost*) is "not wide-awake reality"; its markings are not the same (p. 26).

Kanner, who identified the Pandora, offers a minute description of the wings of the Whites, the antithesis of Vadim's casual, faulty description. Kanner's presence also serves to disconcert the reader, who, by this time, has picked out enough detail to identify the ultimate authorial control of the book only to find a second authorial identity. By this means the reader is parodied.

"Look at the harlequin," Vadim says, pointing to a butterfly: "a smiling red with yellow intervals between black blotches, a row of blue crescents ran along the inside of the toothed wing margins . . . glistening sweep of bronzy silk coming down on both sides of the beastie's body . . . ." (p. 108). This time he has perfectly, though hardly scientifically, described the Small Tortoiseshell of Europe.

"*Krapivnitsa,*" Annette notes, identifying it and the larval food plant by a common designation. Annette was more nettle-ish than nettlefly-ish, and Vadim never succeeded in taming that species.

That there is no butterfly recorded in Vadim's relationship with Louise indicates the prosaic nature of that marriage and that perhaps in it there was no potential either for metamorphosis or transcendence.

The fourth love, identified only as "you," is associated with a nonspecific "yellow butterfly" that settles briefly on a cloverhead and then "wheels away in the wind," like the slip of yellow paper blowing away that occasioned their first speaking to one another (pp. 225-26).

"*Metamorphoza,*" she said in her "lovely, elegant Russian." Elegant Russian it may be, but the comment is absurd, trite, and hardly more than a word-association. However, the idea parallels Vadim's "literary metamorphosis" (p. 122) of changing from Russian to English.[16]

The allusions reinforce the major theme of the book, its authorship and the identity of V. V. Nabokov parodies his own self-conscious use of butterflies, the egocentric literary artist as well as the reader as "triple" harlequins. There is, nevertheless, transcendence through art implied in these harlequins.

In the discussion of these eight novels, we have seen some uses of specific butterflies and some that are not named or identified, which for want of a better term, we shall call "general" butterflies, though even they have traits that place them in definite families. When species are used, they are selected for qualities implicit in the insects themselves, the coloration and patterns, behavior, or life histories that relate to the characters, to the incidents, or else, like Martin's zebra-

27

striped swallowtail, are structural elements in the patterning of the novel. Flight behavior, such as gliding or languorous flight, or darting or swiftly swirling flight, is characteristic of species or of whole genera and are used to convey moods and themes. Martin's blue butterflies are typically settling on damp spots in the road ("tippling" as Nabokov has said more than once), and yet they are "general" because the idea is too ethereal, too nebulous, or too conventional to be precisely defined. So the dark butterflies and Albinus's white butterflies (certainly ubiquitous pierids) in the garden indicate that neither concepts of fate nor certain kinds of eros should be pinned and labeled. When Nabokov uses either a general or a specific butterfly, it is never a casual choice for poetic imagery alone. Anything that remotely suggests a specific concept to be identified by the reader is at least partially defined by a species.

3

Species and Specifics

Unlike the general, non-specific moths of *Despair, The Defense,* and *Laughter in the Dark,* the moths of *Invitation to a Beheading*[1] and *Bend Sinister*[2] are recognizable species, precisely described, although not named. They are structurally important to the novels.

The large moth that appears in *Invitation to a Beheading* is predicted, one of a series, a pattern throughout the novel. Rodion has been feeding insects to the spider in Cincinnatus's cell each morning when he brings the prisoner's breakfast. His fingers had "colored powder" (that is, scales) from a butterfly's wings (p. 124). The spider had "sucked dry a small downy moth with marbled forewings" (p. 169), which could be any one of a number of species. The marvelous moth is that captured on the windowpane of the tower and which Rodion has also brought the spider (p. 203). Accidently Rodion releases it; it lands on his cuff and crawls up his sleeve. In panic, Rodion shakes it off, and it flies to concealment. (The moth asks why it has been disturbed; day and night are reversed for it. Cincinnatus thinks through and for the moth.) Of course, Rodion could not locate the moth. Of course, Cincinnatus knew exactly where it was. This species, the Great Peacock Moth, also called the Greater Emperor Moth, is the largest of European species, with a wingspan of up to 157mm. Its four "eyespots," one in each wing, are basic to Nabokov's imagery involving Lepidoptera, illustrating natural deception, that what is evident is not necessarily real, and also having aesthetic implications. For the text, the size of the moth is important, paralleling the magnitude of the realization that Cincinnatus is about to make. "Everything has fallen into place" (p. 204), like the moth, and he has found the "crack in life," but suddenly his thoughts tangle, and he becomes inarticulate, unsuccessful in expressing the meaning of life. "Death . . . ." he writes and cannot improve on his choice of words. Cincinnatus gets up and, overcome by the invulnerability, the inviolability, and the symmetry of the moth, in his world where everything had been the opposite, he strokes it. In a physical as well as an intellectual sense, small frail Cincinnatus links himself to and identifies with the moth. As he thinks (does not say) later in the disintegration of his cell, all of him will be destroyed and the moth will fly away. The reader can be sure that the moth did escape, as Cincinnatus himself escapes from the prison of himself.

There are also the remnants of a specific butterfly in *Invitation to a Beheading*. Like the "pathetic wing in a spider's web" that Nabokov found in the Pyrenées-Ariège—an unrelated species—this was the "orphaned hind wing" of the Large Tortoiseshell (p. 119). The fact that the species occurs primarily in southern and central Europe and that the wing was in the southwest corner of the web comes closer to establishing a locale for this placeless, timeless novel than anything else in the book.

Nabokov says that the "evil-minded reader" will perceive in little Emmy of this novel a sister to his more famous nymphet, Lolita.[3] Actually ballerina Emmy, more than Lolita, in many ways personifies a butterfly. The terms in which the author depicts her dartings, her flittings, the wing-like spread of her moire sash, her flashing of color are precisely those that he uses for his Lepidoptera. Additional hints are in the manner in which "enigmatically she kept following winding paths with her fingers" (p. 77) like gravid females of some species of butterflies and moths, and the vanilla scent of her hair, which, like Lolita's musky scent, Nabokov attributes to butterflies (*Speak, Memory*,[4] p. 138; *The Gift*,[5] p. 122). Personification is not, however, a device that Nabokov uses (though a few characters sometimes do), and I think neither of these nymphets should be so interpreted.

In *Bend Sinister*, Olga's moth is identified as a sphingid, a "hawkmoth" by the narrator (p. 134). There is no indication that she actually ever held it. Krug, who has already revealed affinity for "the perfection of non-human creatures—birds, young dogs, moths asleep . . ." (p. 27) is visualizing Olga tenderly carrying the moth in her cupped hands and later returning it to the garden where she had found it. It is described in terms of softness: "fluffy feet," "mouse-grey," connoting the tenderness of lovers. The blue eyespots complete the positive identification as the Eyed Hawkmoth, one of the most beautiful moths of Europe. (The iris of the eye is not broken by a black bar as it is in most ocellation-patterns, thus making the eye particularly striking.) Krug was to note a representation of this moth in a nineteenth-century insect book. Characteristically he would relate this moth to Olga and also the caterpillar to his son. The identity of sex and the pink moth is repeated with different implication in his relationship with Marietta.

"This is the pink moth clinging" (p. 197). In the copulation of the Eyed Hawkmoth, the female clings head up to a twig or other solid object, while the less strong male hangs upon the female, legs pressed against the body and its head downward.

The association of hawkmoths (Sphingidae) of various species with sex, often extra-marital, in Nabokov's novels is also indicated in *Speak, Memory*. On a walk with Ordo, his spelling master, Nabokov had found two freshly emerged Amur Hawkmoths, "lovely, velvety, purplish-grey creatures in tranquil copulation" (p. 156). Later he came upon Ordo on his knees, wringing his hands before his astonished mother. The hawkmoths used are those of pink, lavender, and purplish coloration.

The circle is closed when Krug dies (in style) and steps out of the narrative to find the Eyed Hawkmoth striking the window screen and then disappearing into the night. The author-lepidopterist, too, has recognized the "twang" and its meaning for the pursuer. "It is a good night for mothing."

In addition to this important moth there are a number of Lepidoptera metaphors worthy of note, such as Krug's bow-tie with its "interneural macules and a crippled hind-wing," (p. 47), which, with the note to the typist Isabella, establishes the authorship as not merely that of Adam Krug. There are also a number of formic allusions in this work (and in other novels) which are less easily explained.

Velvet, Rowe finds,[6] becomes associated with erotic love or sex in Nabokov's fiction. The hawkmoth is velvety in texture, and the Camberwell Beauty (in America, the Mourning Cloak) is likewise noted for its velvety wings. It is this aspect of the butterfly rather than its essential blackness (deep brown in reality) enhanced by creamy borders that Nabokov uses in symbolizing love, but an evanescent love. In *The Real Life of Sebastian Knight*,[7] the butterfly enters the fourth scenario of the play of changes when Sebastian is with his first adolescent love and they are meeting for the last time (p. 139). Similarly in *Speak, Memory,* this butterfly is associated with Tamara, one of Nabokov's young romances. The butterflies flit through in the autumn of their love (p. 231) and return in the spring, "exactly as old as our romance" (p. 239), and by fall the young couple were to part forever. Hibernating through the winter, the butterfly is one of the earliest to appear in the spring within its range. Nabokov used this species only once in his American novels (*Ada*, p. 170).[8]

With the mention of H. F. Stainton (p. 11), cousin of Virginia Knight, Sebastian's mother, one might expect more lepidopteran allusions.[9] The name of the cousin is borrowed from the English lepidopterist H. T. Stainton, author of *A Manual of British Butterflies and Moths* (1857-59) and founder of the first British journal devoted to entomology. Virginia's migratory flittings, butterfly-like, across southern France and her death in 1909 offer some interesting Lepidoptera associations. In 1909 T. A. Chapman described from the south of France *Callophrys avis* as being a species distinct from *Callophrys rubi*, a species which Stainton would have known in England. Virginia's death from Lehmann's disease suggests, however, not a discovery but the invalidation of a taxon or of taxa such as were involved in the subdivision of the various tribes of *Lycaeides/Plebejdii* (Polyommatinae) by J. W. Tutt and Dr. Chapman in their 1909 papers and which Nabokov discussed in his 1945 *Psyche* article. All of these butterflies are Lycaenidae, an additional interest in view of Nabokov's professional research.

Other imagery in *The Real Life of Sebastian Knight* includes the anatomically impossible butterfly depicted on Sebastian's album (p. 39), Sebastian "freshly emerged from the carved chrysalid of Cambridge" (the words of the biographer Goodman, p. 62), and a simile using side whiskers and the Large Copper as metaphor for extinction (p. 43).

In this novel only one species is named, the Camberwell Beauty, a Holarctic species, and the setting is England and the Continent. In *Pnin*,[10] the locale is the United States, and the butterflies are Nearctic, cosmopolitan, and one which reverts to the Palearctic origins of Pnin and his narrator.

As pointed out in Chapter 1 of this study, *Pnin* records an important butterfly in Nabokov's experiences as a lepidopterist. It is not, however, Timofey Pnin, but Chateau, who points out the blue butterflies (the Karner Blues) that Nabokov had described nor do they particularly impress Pnin. If the blue butterflies represent the kind of joy that Martin in *Glory* felt, in this incident it is the reflected joy of the author, though possibly among his fellow emigres at the Pines, eating borscht and discussing Tolstoy, Pnin was as close to happiness as he might ever achieve. The butterflies serve to pinpoint the locale (upstate New York) and also to reintroduce (Sirin has already been identified among the guests) the author Vladimir Vladimirovich, entomologist (p. 128).

In a later scene, Timofey Pnin does notice the butterflies, though: "huge, amber-brown Monarch butterflies" (p. 136), drifting south. This fall of 1954 was like any other on campus, the students with great ingenuity doing exactly what students have been doing for centuries, defacing ornaments and library books; the perennial parking problems of American campuses had been discussed; the life of academia was repeating its patterns. So, too, the Monarchs were migrating, noticeable in their multitudes as they are every fall. Like Dreyer's Red Admiral (*King, Queen, Knave*), the Monarchs may have aesthetic and some other transcendental value for Pnin. They also suggest the fluctuations in his own life.

The most significant butterfly in this novel, however, is revealed by Pnin's friend and narrator. In the passage he takes us back to St. Petersburg and manages to give a clue to his own character (his "concentrated ecstasy") and to his appraisal of Pnin as an "exceptionally rare aberration" (p. 177). The northern and central European butterfly that he was spreading and pinning, as he was later to "pin" Pnin in his narration, was the "Paphia Fritillary," (the Silver-washed Fritillary), a species of the genus that is featured in *Ada*. The importance of this insect is more in the aberration than in the species, the fusing of the silver stripes being, in Bader's words, a "pattern broken by life."[11] The allusion to the Paphia goddess, Aphodite, should not be discounted; the doctor was one of Liza's lovers.

This "quiet, lacy-winged little green insect" (p. 171) may indeed be Nabokov's intrusion to indicate that under his control the bowl had not been broken,[12] but the insect is more likely a lace-wing (*Chrysiopa*) than a moth.

4

The Gift

The second part of *The Gift*[1] is an exultation of butterflies and a delight to the lepidopterist even more than *Speak, Memory,* which was written many years later. In the latter book, Nabokov acknowledges auto-plagiarism, and the autobiographical theme of the novel lends itself to constant comparisons between the two books. I believe that a lepidoptera-attentive reading of *The Gift* offers the possibility of an interpretation of Fyodor and his father that is not otherwise apparent. At the outset, however, I must emphasize that there is no suggestion of a parallel in the lives of Konstantin Kirillovich Godunov-Cherdyntsev and that of Vladimir Dmitrievich Nabokov other than that they both pursued a goal that transcends the ordinary, though Godunov's was a dual and a clandestine pursuit. The Lepidoptera analogies from *Speak, Memory*[2] are added only in the interest of demonstrating the use of the Lepidoptera in fiction and in the autobiography.

Fyodor had not included for publication his poem on butterflies in his collection that appeared in emigre Berlin because it was too closely related to his father; the time was not yet ripe. The poem (p. 36) is about the first butterflies appearing after the snows have begun to melt from St. Petersburg streets: the Vanessas and the Brimstones, brightly colored butterflies that emerge from hibernation (like the "first Brimstone" of spring in *Speak, Memory,* p. 111). The poem also predicts a divulgence of something less obvious than springtime in the moth image.

> I shall not fail to detect
> The four lovely gauze wings
> Of the softest Geometrid moth in the world
> spread flat on a mottled pale birchtrunk.

Its cryptic coloration conceals the moth. Fyodor was not then aware of the hidden allusion, but this becomes a clue to the reader. Fyodor is shortly to put together pieces of the past and discern a secret life of his father, which finally will cause him to discontinue his research and eventually to write the Cherneshevsky biography instead.

Seeing a rainbow in Berlin reminds Fyodor that his father in Ordos (Shenshi Province, China) had once stepped into a rainbow (p. 90).[3] Immediately after recalling this, a butterfly crosses Fyodor's path, a Poplar Admiral, which is a symbol of his father even to the suggestion that his father, missing now for seven years, indeed has died, an image that is followed by another, that of broken specimens, four wings without body, that one occasionally finds in the woods. The species, and especially the Russian subspecies, had special significance to Nabokov, for his own father had once caught for him a "rare and magnificent female" (*Speak, Memory,* p. 192, also p. 133).

Fyodor then begins a walk into the past, recalling his mother and father sitting on a bench on the eve of his father's departure for butterfly-hunting expeditions; the past fades, leaving a scrap of rainbow, a butterfly in disintegration—three wings, no abdomen—on a pin, and a carnation lying in the sand, and Fyodor returns to the present. When his mother is visiting him in Berlin, she wonders about Godunov, possibly still alive in Tibet or China, but perhaps not. In imagining his father's sudden return and an explanation of his absence, Fyodor can feel only a "sickening terror, not as of a ghost, but of one that would not be frightening" (pp. 99-100). On the evening before her return home, she recalls the lepidopteran verses that Fyodor and his father used to compose. They speak of the *Epicnaptera* moth,[4] which reflects in its name the epic quality of the father's life, though not necessarily either to Fyodor or to his mother. It was while reading Pushkin's *Journey to Arzrum*[5] and a passage about frontiers that Fyodor felt a sudden stab and immediately conceived the idea of his father's biography, though he divulges his plan to his mother only by letter. She warns him that he needs facts, not family sentimentality, and gives hints about whom he should consult for his scientific work. Fyodor immersed himself in Pushkin, his father's favorite poet and sees a Niobe Fritillary and a small Black Apollo (p. 120). The reader must be alert not only to Pushkin's beautiful lines but also to Niobe weeping over her children killed by Apollo and Artemis. It is another clue that the author brings out through poetry and butterfly allusion.

Fyodor recounts the facts of his father's life: his writings on Lepidoptera and "The Travels of a Naturalist," and an article on the geometrid mimicking a small parnassian, patterning back to Fyodor's poem and the Apollo allusion in Lepidoptera references and his travels into Asia, including China. All of these facts Fyodor took from an encyclopedia, insufficient material for a biography. In reply to his query for something about their life together, his mother tells him of his father's devotion to his Lepidoptera and how he had sent her back home when she had followed him to the camp of his expedition. She actually reveals nothing of their conjugal life. To this point, Fyodor had not discerned his father's secret, and his research had been happy.

Fyodor recalls the "Paradise smell" of the laboratory and the family members who talked of butterflies as if they existed only in so far as his father existed (p. 118). Fyodor, too, had fallen under the spell of butterflies (not the *passio et morbo* of Nabokov), and his father instructs him but does not consent to take him on the expeditions. Fyodor imagines the expeditions of his father as if he were a participant. And yet, bad in geography in school, now that he is working on the biography, he fears an "idiotic blunder" in describing his father's distant research. Another hint for the reader but not yet for Fyodor.

The recollection of Godunov's catching his first Peacock Butterfly (p. 121) reminds us that Nabokov's father also recalled the capture of this species and also near a little bridge (*Speak, Memory,* p. 75). Fyodor's experience with his father on the use of the microscope is also echoed in Nabokov's autobiography, but in remembering it, Fyodor writes, "Oh, don't look at me, my childhood, with such big frightened eyes" (p. 121). The truth of something he would prefer not to admit is becoming clearer.

The description of Godunov's lessons to his son center around references to procreation, which Fyodor mixes with non-sexual observations, swinging from natural scenes to laboratory, from summer to winter, from cold to heat: the genital armature of butterflies, the artificial creation of aberrations in the imagoes by alternating the exposure of the pupae to heat and cold, the lycaenid-ant symbiosis, the odors and the sounds ("voice"—actually various clicks, squeaks, etc. produced by caterpillars, pupae, and butterflies of certain species by motion of parts of their bodies), mimicry and migration. Of immense pride was Godunov's discovery (simultaneously with that of Tutt)[6] of the corneal formation (the sphragus) of female parnassians. Fyodor feels that he can preserve this part of his father's life, but he remembers how his father possessed a secret knowlege of something that had "no direct connection either with us, or with my mother, or

37

with externals of life, or even with butterflies . . ." (p. 126); there was something about his solitude, as though he were fleeing from something, and about his enigmatic personality that Fyodor cannot understand. This aspect of Fyodor's father suggests transcendence, but it is paradoxical in that it is not, apparently, the forward movement such as would be implied in artistic-scientific pursuit.

Fyodor knows that Godunov left his party for days at a time, presumably alone, and he wonders if at those times he ever thought of his wife and children.

Fyodor reconstructs a journey with his father through China with its marvelous flora and fauna and especially the remarkable Lepidoptera in all stages of metamorphosis, imagining that he had been on that expedition. He then recalls his father's last return and the rare butterfly, its body still not firm on the spreading board, that he had collected and was eager to show him. Shortly after this, the family had travelled to the Crimea. Suddenly, restless and unhappy, Godunov packed up and left for the trip from which he never returned, even recognizing the criticism that he would engender by having left his country in the middle of its war with Germany. Moments after his departure, a "familiar black and white butterfly" (the Poplar Admiral, his father's symbol) appeared, and Fyodor went into the meadow, the "divine meaning of which is expressed in its butterflies . . .. But its truth would have been probed somewhat deeper by knowledge-amplified love" (p. 144).

The butterflies that Fyodor sees in this meadow reveal something of the secret life of his father: "merry-looking Selenes" (fritillaries) express his father's joy at departure; they are also at the height of their mating season; the Swallowtail, like his father on those expeditions "bedraggled but still powerful"; the Black-veined Whites, stained with fatidic pupal discharge; the Ringlets and the Burnet moth, metaphors of other customs and costumes; the gravid Cabbage Butterfly; the male Coppers (*their* females had not yet emerged from winter diapause but as soon as they appeared, mating would take place); the Freya Fritillary with an evocation of the Goddess of Love; and the mimetic humming-bird moth. All are images of love, family life, and disguise. Fyodor leans against the birch tree and bursts into tears. He had been left behind again; in the meadow he also realizes something about himself and an inkling of something else about his father which is to become clear only seven years later as he progresses on the biography.

The last report of his father had come from a French missionary;[7] the two of them had spent time discussing the name of a light blue iris:

38

a pattern from the earlier mention of him riding, in Fyodor's imagination, "across a vernal plain all blue with iris" (p. 33), and from Fyodor's wallpaper and his Berlin landlady's dress, both yellow with blue iris. The pattern will be repeated in the ceiling painted with Tibetan butterflies (p. 366), a significant allusion.

Continuing his imaginative recounting of his father's trip through Asia, he conjectures on the possibility of his being alive in China or if dead, how he might have died. Even at the point of execution, Fyodor is sure he would have leaped out to follow a white moth as he had the pink hawkmoth "sampling our lilacs" in Leshino (p. 149). The pink hawkmoth suggests that Fyodor has now realized the truth of his father's life: that his wanderings, sudden departures, long absences, and strange behavior are indicative of the double life of his father, with a mistress, possibly even a second family, in the Far East. (A second family is suggested in the gravid butterfly and in Fyodor's searching for the larvae of the Imperatorial Apollo on his imaginary trip through Tatsuin Hu, p. 135.) A hint of Godunov's liaison in China or Tibet was dropped on the first page of the novel in the couple that Fyodor saw on the Berlin street: "he, beetle-browed and old; she, no longer young, with a rather attractive pseudo-Chinese face" (p. 15).

Later Fyodor was also to watch a geometrid caterpillar, the larvae of his father's "soft secret" in Fyodor's poem.

Having realized this secret life, Fyodor abandons his project and even writes his mother that he will read part of it to her (not all, we note), that he fears contaminating the image of his father's travels, departing from the real poetry with which "these receptive, knowledge-able, and chaste naturalists endowed their research" (p. 151). He chooses to maintain the deception to his mother, who no doubt would catch the irony of his statement but in any case is perfectly aware of Godunov-Cherdyntsev's duplicity.

Nature's ingenious deceptions are again remembered when Fyodor leads the reader into the Grünewald forest, and an Anglewing, a leaf-mimic, suddenly appears (p. 344). The species also appears, "coming from nowhere" in *Speak, Memory* (p. 106). The "white bracket," or initial (a comma or C) on its lower hindwings was clearly visible. Though this was near the spot where Yasha Chernyshevsky had taken his life, and the initial is correct, Fyodor does not recognize a symbol; he is concentrating on his perception of art through nature.

The white commas of the Anglewing have a counterpart in the negative image of black commas on the "golden, stumpy little butterfly" (p. 346f). The Silver-spotted Skipper, ranging across Europe and Asia (as well as North America, where it is known as the Comma Skipper), "suddenly alighted on an oak leaf, half-opening its slanting wings and suddenly shot away like a golden fly." Fyodor is reminded of his father's wanderings, and despite the beauty, peace, "something akin to that Asiatic freedom spreading wide on the maps," he forces himself to believe that his father is indeed dead.

Toward the end of the novel, Fyodor concludes: "The most enchanting things in nature and art are based on deception" (p. 376). His father's life is described in dual aspect; the revelation of Godunov's secret no doubt influenced his composition of the biography of Chernyshevsky.

5

Pale Fire

The Lepidoptera of *Pale Fire*[1] have major roles in the basic patterning of the novel, the "contrapuntal theme" between Shade's poem and Kinbote's commentary,[2] both in serious concepts of death and the hereafter of the poem and in the mock-serious and humorous content that Kinbote provides.

Son of naturalists and influenced by Aunt Maud, John Shade has respect and appreciation of nature and especially of zoologic forms, birds and insects, which to him are images of his art and of a larger meaning of life. He addresses his poem to his wife Sybil, *his* Vanessa, in an effort to explain his concepts of these metaphysical subjects. The moths and butterflies are part of nature's art, a world of transparency and iridescence, reflecting an afterlife in which man is "most artistically caged" (line 114). Metamorphosis is a mystery that correlates to the authorship of life and creativity in the practice of art. Kinbote, on the other hand, not only acknowledges his "limited knowledge of lepidoptera" (Index, p. 308) but fails to understand the metaphors of Shade, and in his "realistic" interpretations of the images draws ludicrous conclusions.

"White butterflies turn lavendar" when they flit under the branches of the hickory tree from which hangs Hazel's swing (lines 55-57). These general butterflies suggest not only the spirit of Hazel but of others as well, and the colors of life beyond death rather than an achromatic non-existence. The Luna moth ("a large, tailed pale green moth," Kinbote quotes from his dictionary, p. 114) Shade has not used for the spirit of Aunt Maud and has also rejected the extended image of the cocoon, "the leaf sarcophagus," as not representing sufficiently her continued life beyond the grave. Kinbote explains the deletion (if it *was* Shade's, not his own emendation) as a clash in the moth's name with "Moon" of the next line. The moth is not a spirit transcendent.

Shade, Kinbote tells us, was always pointing out "natural history" and commenting on the plants and animals. Kinbote was not much interested that the "diana (presumably a flower)" occurs along with the "atlantis (presumably another flower)" in New Wye (p. 169). The Diana, a large beautiful butterfly, is known for its rather spectacular sexual dimorphism. The male has velvety black and orange banded wings, while the female is predominantly black with a wide border of iridescent blue on the hindwings. The Atlantis, also a fritillary, is rich orange with black markings. Both species feed on violets in the larval stage. The reference has a counterpoint in the poem.

Death is put into metaphor as "the Worm" by the director of the Institute of Preparation for the Hereafter, but whether it refers to annelids with burial allusions or to larvae is not made clear. For Shade, it would certainly have been the former, as change and growth are implied in life but not in death or "Elysian life." Shade himself uses a larval image in "life, the woolly caterpillar" (line 667), which Nabokov expanded in *Ada*.

The I.P.H. "was a larvorium and a violet" (line 516). This is Shade's contrapuntal reference to fritillaries and implies the illogical and inconclusive philosophy of the Institute: the larva before the egg has been laid on the larval food plant, and no final metamorphosis.

In erudite John Shade's allusions exist other possibilities that pertain to Lepidoptera. A fritillary fairly common throughout moist areas of temperate North America and ranging into the Rockies, where presumably the Institute was held, is the Cybele (*Speyeria cybele*, the Great-spangled Fritillary). The fact that this name is assonant with Sybil, to whom the poem is addressed, may have little significance, though it is in that passage that Shade remarks on her little ways and habits that endear her to him. The festival of the Roman goddess Cybele is celebrated on March 22 with the cutting of a pine tree ("L'if, lifeless tree" line 502)—not a yew, but needle-leafed like the pine and also native to the Rockies. The trunk of the pine tree was then swathed like a corpse and decked with violets ("a grave in Reason's early spring"—line 517).[3] Shade might also be alluding to the Larvae (Lemures), the malificent spirits that visit the house of the dead on certain days in May. Despite the first syllable, with "if" and the initials of the Institute, the European butterfly *Iphiclides podalirius* is not suggested.[4]

Kinbote tries to explain the Toothwart White (p. 183f) by using his dictionary and is only confused by the definition. An alternate nomenclature, "Virginia Whites," Shade had discarded and at the same time changed the verb. Characteristically Kinbote misses Shade's association of the white butterfly with the spirits of the departed.

The most striking of the butterfly images is that of the Vanessa, the Red Admiral or Red Admirable, and the antiphonal theme of Shade's concepts and those of Kinbote is illustrated in the different ways the two writers refer to the butterfly.

When Shade evoked Sybil with

> Come and be worshipped, come and be caressed,
> My dark Vanessa, crimson-barred, my blest
> My admirable butterfly   (lines 269-71)

he was speaking in tenderness, devotion, and admiration for her and the spirit in which she saw life. The Red Admiral symbolizes his love. The velvety aspect of the butterfly aroused him; the colors of the butterfly are those of passion.

For Kinbote, even though it was familar to him in the Palace Gardens of Zembla, the butterfly is important only as an heraldic emblem, "borne in the escutcheon of the Dukes of Payne" (p. 172). He associates it not with love nor with beauty but with "oozy plums and a dead rabbit,"[5] and calls it "a most frolicsome fly." It must be pointed out, however, that in calling it a "fly," he is not being inexact nor disparaging but only archaic, as the term was used for butterfly in the eighteenth century and even later. This pretender-realist could have meant it either way, of course.

Regarding his use of the Red Admiral, Nabokov said it was familiar to him in Russia, where it was called the "Butterfly of Doom," because it was especially abundant in 1881, the year Tsar Alexander II was assassinated, and the markings on the underside of its two hindwings seem to read "1881."[6]

Just before his death, Shade sees the Red Admiral as it "wheels in the low sun, settles on sand/And shows its ink-blue wingtips flecked with white" (lines 993-95). Though it may presage his death, it is still for him a creature of beauty, a wonder, and it is accurately depicted in his poem. Shade had premonitions of his death; he may even have recognized it as symbol.

In contrast, Kinbote sees the same butterfly as it "came dizzily whirling around us like a colored flame." He comments more on the sun and the sand and becomes quite humorous in describing the butterfly's ecstasy and frivolity but misses, even in retrospect, the significance of its appearance at that point in Shade's life. He does, however, finally describe its dissolution in the shadows, "a magnificent velvet-and-flame creature" (p. 290). He does not see the connection but unconsciously has absorbed part of Shade's appreciation of it.

One of Kinbote's best lines also relates to this species. In the Index, he describes himself "with the sable gloom of his nature marked like a dark Vanessa with gay flashes" (p. 309). There is no such reference on the page indicated, 270, but it is anything but a "blind" reference.

In his own story, Kinbote uses a black butterfly, which like the one Martin (*Glory*) saw on the mountain, suggests the intrusion of fate. The scene in the Zemblan mountains has an eerie setting of sheer cliffs in the "sepulchral chill" of early dawn. With the first rays of the sun, the black butterfly "came dancing down a pebbly rake" (p. 142). The fate symbol is shortly dispelled, however, as danger passes, and the girl is about to disrobe for King Charles's pleasure. Kinbote is too obtuse to glimpse meaning of any kind, fatidic or sexual, in the butterfly. There is an element of burlesque in the passage, though, because the butterfly most certainly was a member of the family Satyridae, and King Charles was anything but amorous of nymphs.

6

Lolita[1]

In his introduction to the *Annotated Lolita,* Alfred Appel asserts that the novel is too "transcendent" to have its hermetic components worked out to form a meaningful pattern and that an attempt to do so becomes a parody in itself.[2] Nevertheless, to follow every lead in a novel with a Lepidoptera-centered theme is a temptation that a lepidopterist could hardly resist, and Appel is very right in concluding that much of this attempt at exegesis comes to a dead end.

Lolita is not a personification of a butterfly, though, as Diane Butler points out,[3] many of her motions and her actions are analogous to those of a butterfly. The epithet "nymph" is, I suggest, a part of the author's disingenuousness. As a metaphor for youth, it has no reference to Lepidoptera, and yet the word carries not only the appeal of mythologic femininity but also the charm of the wood and grass nymphs of Satyridae (also an enthusiasm of Nabokov as reflected in his 1942 article) and of Nymphalidae. Primarily, though, the term is used for the young of those orders (Odonata, Hemiptera, etc.) which, upon hatching from the egg, resemble the adult it will become, unlike Lepidoptera and other orders which in larval stage in no way resemble the adult and pupate for complete metamorphosis to adult insect. Lolita is very young when she enters an adult sex life, and her imagery could be as well that of damselfly or dragonfly (Odonata).

It is, of course, not Nabokov but Humbert Humbert who calls Lolita a nymph or nymphet, and there is scarcely an entomological reference that he uses accurately. However, when he says that "nymphets do not occur in the polar regions" (p. 35), he is quite correct. There are no odonates—but there are butterflies within the Arctic Circle. A further support for this imagery is in *Look at the Harlequins!* when V. V. speaks of his novel of a "libellula girl on an elephant" (p. 228) (Libellulidae: Odonata). Whether or not the dragonfly imagery is inherent in the novel, *Lolita* is based on the *pursuit* of a butterfly for scientific description, and the butterfly hunt was for a lycaenid, not a nymphalid, or satyrid.

That metamorphosis is present, especially in the developing love of Humbert Humbert for his nymphet could hardly be denied, but it is not metamorphosis in entomological terms. There is no development that could correspond to instaric lives and pupation-eclosion in *Lolita*.

The introduction by the pre-Linnaean systematist, John Ray (Jr.) and the dates of 1952 that relate to Humbert's crime reveal the theme of the novel. Nabokov in his afterword to the American edition picks ten passages for his "special delectation" some of which are patently lepidopteran in reference, but he also gives hints in the course of that essay to the novel's position in relation to his activity in Lepidoptera (beyond the "mimic and model" association of John Ray, Jr. to the author).

The action, for a pornographic novel, he says, must consist of a "copulation of cliches"; it must have an "alternation of sexual scenes," a reduction of connecting passages to the "sutures of sense," "logical bridges of the simplest design," and "brief expositions and explanations." Standard terminology, an examination of the genitalia with detailed descriptions and explanations, the "logical bridges"—these are the schema for his scientific publications in Lepidoptera, even to the "lewd lore" at the conclusions (diagrams of genital armature). There is no question of applying the term "pornography" to *Lolita*; Nabokov is being both revealing and deceptive. Furthermore, in the essay, he uses other metaphors such as "the nerves of the novel," and "among discarded limbs and unfinished torsos" to corroborate the laboratory perspective of the theme. The controlling metaphor is not, then, a butterfly, though assuredly Lolita/Dolores/Dolly ("doll" from the Latin *pupa*) carries the weight of the butterly association in name. Rather, the underlying theme is Lepidoptera activity from schematic classification and the description of a new species to the finding of the female *Lycaeides a. sublivens* Nabokov near Dolores, Colo. That Humbert Humbert was apprehended for his "crime" in September 1952, the month after Nabokov wrote of his finding the female *sublivens,* is not as important to the text of the novel as it was to the author. Cognoscento John Ray, Jr. knew that fact, as few of the readers of the novel would. I suggest that Humbert also manages by verbal contrivance to reveal authorial/*auctorus* identification, however briefly, in Taxovich (Maximovich) as taxonomist.

When Humbert Humbert says that he has control of legalistic

matters (i.e., the code of zoological nomenclature) and then lops off the *nomos* to substitute the patronymic ending in referring to this White Russian, he is indicating another kind of family relationship and identifying the author. Humbert is not the scientist himself (quite the contrary!), but by his extraordinary means, including ignorance of the subject, he brings the focus back to the author at this and several other points in the novel. Taxovich consults Humbert about Valeria on subjects that in Lepidoptera research would be preliminary to the study of a species: her diet (the larval food plant, since it is on these that the female insects will be found, ovipositing); her periods (that is, the instars); her wardrobe (the colors and patterns of the butterfly that distinguish it). The "books she had read" might refer to previous sources that described the species or related ones (in this case Nabokov's own MCZ monograph), but it also disguises the parallels by putting the passage into human context.

The Ramsdale School list is a natural for a list of taxa, but whether the forty names (or thirty-nine, excluding McFate)[4] represents a list of genera or of species of certain Lycaenidae is problematic. While there is a Falter (German for butterfly) and the surnames of several lepidopterists, there are also literary and artistic names that connote disguise, deception, and enigma.

The term "waterproof" (p. 91, p. 274) could refer to the nymphs of the odonates or a projection of the nymph-Lepidoptera imagery in reference to the *Nymphulinae* larvae (moths—microlepidoptera), one of the few aquatic Lepidoptera in larval state.

The "faint violet whiff and dead-leaf echo" (p. 279) of seventeen-year-old, pregnant Lolita suggests butterfly mimicry, the scent of certain butterflies, and the food plant of fritillary larvae.[5] However, the image is more closely linked, it seems to me, with the possible fading (i.e., the loss of minute scales) of the colors of the type specimens even in the best-kept collections.

The Elphinstone hospital scene was another of Nabokov's chosen ten. There is a genus *Elphinstonia* (Pieridae) with no North American representatives, however, and the elfins, such as the Bog Elfin, are in various genera of Lycaenidae. I think, however, that these are too tenuous to be included as elucidation of this passage.

Most of the actual butterfly allusions in *Lolita* Nabokov explains to Alfred Appel and are included in *The Annotated Lolita*, "Notes." Since Humbert Humbert is a parody of persons with superficial interests in natural history (and the book a parody of the careless reader as well as the exegetical one)[6] and since Humbert knows little

47

about insects and is incapable even of using a field guide to identify wildflowers (p. 246), Nabokov's explanations are most helpful to the more-than-superficial reader who *does* care about those insects.

Just before he notices the stranger on the porch (a shadowed Quilty), Humbert sees the "hundreds of powdered bugs" around the light (p. 128), and Nabokov says that these are "millers," noctuids and other moths.[7] The "creepy white flies" around the yucca (p. 158), Nabokov identifies as *Pronuba* (Prodoxidae),[8] the yucca moths in an interdependent relationship to the plant, and Humbert's "gray hummingbirds in the dusk, probing the throats of flowers" (p. 159) as hawkmoths.[9] Nabokov repeatedly uses moths around lamps to indicate love deceived and hawkmoths as symbols of sexual passion. He had, by the way, cited both the Broad-Tailed Hummingbird and the White-Striped Hawkmoth as "patronizing" the gentians in his article on the female *sublivens*.[10] "Hummingbird Moth" is equally valid in popular nomenclature.

Nabokov also explains how he has used names of butterflies and lepidopterists in personal names of some of his marginal characters. The sisters Edusa (p. 211) and Electra Gold (p. 233) are the pierids *Colias edusa* (i.e., *C. crocea*) and *Colias electra*,[11] names that are no longer accepted as species designators. Avis Chapman is formed from the species and its discoverer, omitting the genus of the south European butterfly, *Callophrys avis* Chapman.[12] The species is a hairstreak (Lycaenidae) named during Nabokov's youth (1909); he mentions it in *Speak, Memory* as one newly described that he sought in a Berlin butterfly shop (p. 205). Avis was possibly a female in Dr. Chapman's life and has nothing to do with birds but orders the butterfly back in proper phyllogenetic perspective (Insecta, Aves, Mammalia).

On other points Nabokov appears to be deliberately elusive if not downright misleading, playing games with his critics as well as with his readers.

When Humbert is watching Lolita pay tennis, suddenly, in a single line paragraph:

"An inquisitive butterfly passed, dipping, between us" (p. 236).

It is true that butterflies are inquisitive, and the dipping motion is characteristic of a number of genera, but Nabokov denies symbolism.[13] Still, the "absurd intruder" that appears while Humbert is on the telephone is like the Boschean butterflies (also used in *Ada*), and "he waved his wrists and elbows in would-be comical imitation of rudimentary wings" (p. 237). Quilty does come between Lolita and Humbert.

Nabokov also manages to offer an additional chuckle in his explanation of Rita's use of the simile of the drive around Grainball City, "going round and round us like God-damn mulberry moths" (p. 261). He explains the reference to the Bombycidae, or silk moths of China.[14] What he does not explain (expects his readers to know?) is that Rita's simile is most inept, since these particular silk moths, through breeding for commericial purposes, have long since lost the power of flight. The final laugh is again Nabokov's.

The white moths that drift into the headlights of Humbert's car (pp. 294, 296) may be, like the white moths of *Laughter in the Dark* symbols of deceived loves, or they may be the spirits or spectres (as Rowe would have it)[15] of the departed. That they are also the signature of the omniscient author is undeniable.

Throughout *Lolita* there are numbers and series of numbers that have baffled critics and exegesists. Some of these, like the figure 8 on Lolita's arm (p. 236) may be catalog numbers for specimens or biostatistics of one kind or another, not necessarily related to *sublivens*, male or female at Harvard, Cornell, or the American Museum of Natural History. There is also concealed comment on scientific method in systematics and on not-so-scientific method in natural history.

Quilty tantalized Humbert with his fictitious names compounded of multifarious allusions which he signed on hotel registers, including "Morris Schmetterling," author of "*L'Oiseau ivre*" (p. 252). The farcical element of this confused authorship is repeated when Quilty, with a gun pointed at him, offers Humbert all his property and says that he has been called "the American Maeterlinck. Maeterlinck-Schmetterling " (p. 303).

This scientific-literary association is a statement on deducing from observation of natural history tenets that are not scientific. Maeterlinck in his writings on social Hymenoptera (bees, ants) and Isoptera (termites) reached moral conclusions regarding the lives of the insects. By using the German word for butterfly, Nabokov (not Quilty) puts in the same category the various practices of species description that he found faulty among some nineteeth-century lepidopterists (cf. *Speak, Memory*, p. 123f) and the untenable entomological conclusions of Maeterlinck, at the same time suggesting that Quilty as a playwright is also farcical. He may also suggest in Mrs. Vibressa the controversy between geneticists and morphologists ("bristle-counters") in species differentiation (in Lepidoptera, it would be scale-counter) in the science of systematics.[16]

7

Ada

Ada is about incest and Time, and like all of Nabokov's novels, artistic endeavor—and more. It is also typically spiral, the loops or "twirls" of patterns, themes and imagery, and the author's identity is established in references to his other novels and in a special butterfly. Science (insects) and incest are carefully interwebbed, at times coiled, to create one of the principal images that disguises the real essence of a Texture of Time which is the novel *Ada*. The transcendent butterfly and the sex-connoting hawkmoth are present in *Ada* and other butterflies and moths, but unlike the lepidoptera of his other novels, the imagery is largely larval, and the texture of caterpillars is equally important for Time and for the erotic sensation which is its most obvious function in the novel.

In *Pale Fire,* John Shade used the image that is elaborated in *Ada*[1]: "life, the woolly caterpillar" which parallels Van Veen's "still soft, long larval 'now' " (p. 539). Time is but memory in the making; a sense of Time is a sense of continuous becoming, Van concludes (p. 559). He has wished "to caress Time," "the coolness of its continuum" (p. 537), as he has caressed larvae, actual and invented—as well as Ada, Cordula, and Lucette—throughout the novel. The larval imagery is not exclusively associated with Van, however, nor with his conception of Time; it is used also with reference to Ada, Lucette, and minor characters.

For a better understanding of the text (and texture) of the larval imagery, a few points about Lepidoptera larvae should be noted. (With apologies to those who know very much more about the subject and are aware that treatises can not do justice to it.)

While the species in most cases can be identified by its larva, the larva in no way resembles the mature butterfly or moth. Within the species there is slight variation among individuals, yet in the various instars, some species change greatly. Some larvae are smooth; others have setae, long or short, soft or stiff, simple or spiked depending upon the species. Some are colorful, with or without disruptive patterns, and some have mimetic patterns, such as eyespots, that would startle and deceive a predator, while others are quite plain and blend with their natural surroundings. All are elongated (eruciform), though lycaenids appear slug-shaped. Sphingid moths get their appelation from the sphinx-like position of the heads of their larvae, especially when disturbed. All have the primordia of the wings in form of pouches (invaginations) developing, or everting, only in pupation. Some species are gregarious at least through early instars, while others live solitary lives on their foodplant. While Lepidoptera are vulnerable in all stages of their lives, they are most vulnerable in the larval stage: to disease, accident, starvation through dessication or loss of their foodplant, to parasitism, and to predation. In the normal life cycle, the larvae after a final moult will pupate, some, such as many moths, in cocoons spun from silk from special glands, others in papery cocoons, some without any cocoons at all, and many butterflies in chrysalids suspended and exposed. Hormonal changes in pupation result in the development of all the primordial features that were present in the larva, indeed even in the egg stage, and eventually the imago, or mature insect will emerge.

Shortly after Van arrived at Ardis, Ada explained her fondness for larvae.

> *"Je raffole de tout ce qui rampe* (I'm crazy about everything that crawls)," she said. (p. 54)

Despite this statement, she shows interest not in other forms of creeping zoology but only in caterpillars, and these are a combination of fantastic and authentic species befitting Ardis and Antiterra. The Sharkmoth and the Vaporer Moth (through which Van gains "esthetic empathy") are actual European species. The Puss Moth larva that Ada so expertly described in her youthful notebook is both modest and a diabolic monster "with front segments shaped like bellows and a face resembling the lens of a folding camera" (p. 55)—a suggestion of her much later activity with Lepidoptera—but she also tells of the pleasant sensation of stroking its bloated smooth body: "the sensation is quite silky and pleasant" until it exudes its acrid fluid. The catocala moth, later named, and the Carmen Tortoiseshell are invented species. Her next entry is most extraordinary even for a very precocious twelve-year old.

Having read Proust, she gives a remarkable description and a fanciful name to an actual species:

> ... the noble larva of the Cattleya Hawkmoth ... a seven-inch long colossus, flesh-colored, with turquoise arabesques, rearing its hyacinth head in a stiff Sphinxian attitude. (p. 56)

Her sexual proclivities were as natural to her as the tenderness with which she addressed "cats, caterpillars, pupating puppies" (p. 119).[2] That the Cattleya Hawkmoth is actually the Convolvulus Hawkmoth is evident only in a subsequent passage.

Ada creates in her artwork some equally strange fauna-flora hybrids. Using illustrations from a book (a fusion of art and science), she makes combinations of orchids (some of which are also spurious) to produce a picture of the "marvelous flower that simulated a bright moth that in turn simulated a scarab" (p. 100). She employs a human depiction of nature in art to create mimicries of nature in her own art, which do, in fact, have some counterparts in actual nature. In the context of Ada's and Van's relationship, her drawings also represent her involvement in incest.

When Ada as a pre-adolescent and later in retrospect tells of her plans in Lepidoptera, she reveals two dreams. One is an Institute of Fritillary larvae and their food plants, violets (p. 57). In her plan there may be indication of her desire for natural marriage and procreation.[3] She will breed butterflies and violets, some of them rare, from all over the world. She already knows the techniques for breeding Lepidoptera by holding the male and female in proper position.

Her second plan is for a book depicting the argynninarium and its contents, the fritillaries from egg to pupa, colored plates of each instar of each species, and line drawings of the mature insects' genitalia and other structures (p. 404f). This was to be accomplished in collaboration with Dr. Krolik, but Ada envisioned herself doing the art work and the breeding. Notable in these is her lack of interest in the mature butterfly except for procreative purposes, though she had done some preserving of species for collections upon their emergence from pupation. Her "institute" would be a mini-Eden, a very specialized and in most ways an artificial one, since her Eden would not be a "garden of earthly delights" for the creatures but a beautiful violet-flowered prison for breeding inmates. Her proposed book, while purposeful for science and also a work of art, does not emphasize the butterflies themselves. Her creativity, her concept of metamorphosis is complete only in the strict biological sense; it does not include any aesthetic value for the transcendent beauty of the emerged butterfly. Her art work is also limited to a utilitarian function.

Van had been unable to show much sincere interest in Ada's larvarium, but before he left Ardis that first summer, he observed the metamorphosis of her insects: the sharkmoth in pupation; the Lorelei underwing parasitized and dead—its disguises not deceiving the ichneumon wasp; the Vaporer in pupation; the Puss Moth larvae "ramping" (a natural activity preceding pupation) and the fantastic Carmen already a butterfly but soon to be pinched for preservation. The "Odettian Sphinx" (Ada's Cattleya Hawkmoth) was also in pupation: "an elephantoid mummy with a comically encased trunk of the guermantoid type," more Proustian references describing the pupa of the Convolvulus Hawkmoth with its proboscis case shaped like an elephant trunk.

Meanwhile Dr. Krolik was chasing a special pseudo-pierid, the *Antocharis ada* Krolik (1884), the Ada Orange-tip, with its genus corrupted from the actual *Anthocharis*. This totally fantastic species failed, since one Stümper (German for "blunderer") had described it in 1883, and by the "inexorable law of taxonomic priority," it was *A. prittwitzi* Stümper (p. 57). O. von Prittwitz was an actual nineteenth-century German writer on Lepidoptera.

The Carmen Tortoiseshell and the small Orange-Tip, like the "gitanilla" Lolita and Ada's "lolita" skirt with red flowers on a black background suggest the passion of sex. The caterpillars and the hawkmoths, too, have sexual connotations, and catocalas (the Lorelei, and the *adultera* of *Speak, Memory*) are known for their suggestive names.[4] The Orange-Tip is to have further reference at a later time in Ada's life.

The butterflies of Dr. Krolik require special attention in considering the theme of nature in relation to art. None of his species is an actual, bonafide species; they are all described in terms of gems or jewelry; they all have sexual innuendoes, and they all fail in one way or another. "Inestimable gems" because of the silvery aspect of their wings, the three Kibo fritillaries in their "exquisitely carved chrysalids" that Krolik brought Ada as a birthday gift (p. 79) also have erotic connotations in the shapes of the three mountain peaks of the Mt. Kilimanjaro massive. These, like the Lorelei Underwing, are parasitized, their metamorphosis unachieved. It is possible that Krolik, a physician, had taught Ada more than the reader is permitted to learn (more than Van knew)—if nothing more than some "gem of wisdom" that would affect her attitude toward procreation but possibly something very artificial or abnormal that could have affected her fertility or her ability to bear children. Later the implication is repeated; Ada goes to the gynecologist Seitz, whose name is borrowed from Adalbert Seitz, eminent German lepidopterist.

When Dr. Krolik died, Ada interred her living larvae with him, a symbolic turning point in her life. However, as she and Van edit the book in their old age, she denies the interment of the larvae, saying that she turned the larvae loose on their proper food plants and buried the pupae (that is, not the suspended chrysalids of most butterflies but those of parnassians as well as sphingids and other moths) so that their metamorphosis would be complete (p. 193). By this time she would consider burial of living larvae and pupae with Dr. Krolik as reprehensible.

After his death, their collections of mounted specimens were confiscated "by a regular warren of collatoral Kroliks to agents in Germany and dealers in Tartary" (p. 405).

Krolik is purposely little more than a figurehead in the novel despite the reproductive leporine qualities of his name,[5] an unachieved character whom Van never delineates as a person perhaps because he was a potential threat or because he represents a part of Ada's life that Van never fully participated in. Krolik's total failure in a work of art, as his total failure as a "real" lepidopterist, lies in his unscientific and unaesthetic values. He was Ada's "court jeweler"; he speaks of Lepidoptera neither for their essential place in nature nor for their purpose to science but only as precious gems in a collection. He uses metaphors of inorganic nature for organic nature, suggesting transposed values. His appreciation of nature is only slightly less solipsistic than Van's.

Krolik's artificiality has a counterpart in the aesthetic that Demon defends in considering the Bosch painting (pp. 436-37). To Demon art and science melt "passionately," "incandescently," "incestuously," in a representation, pictorial or emblematic. He refuses to see symbolism, the "myth behind the moth," or any deliberate underlying meaning to Bosch's painting of the uppersides of the butterfly wings rather than the correct undersides considering the position of the butterflies. He sees them as "casual fantasies" for the "fun of contour and color"; the appeal to the senses should be sufficient. Bosch defied science; Demon elevates art above science.

In Antiterra there are authentic species, some of which are represented in art, and invented species. Assuming that they are Nabokov's, not Van's, the fantasy species form a pattern or spiral: art imitating nature which in itself is nature imitating art. ("Nature imitates art," Nabokov says in *Speak, Memory*.) To indicate that he, not nature, and not Van, is in control, he adds a Nabokovian butterfly to the novel.

When Ada and Van have managed to exclude themselves in the park near Ardis for their secretive sex, Ada points out, as Van writes in his memoirs: "some accursed insect that had settled on an aspen trunk" (p. 158). The butterfly conceivably stands as a symbol for their passion, but it is also Nabokov's own creation.

"Accursed? *Accursed?*" Ada's marginal note of many years later continues:

> It was the newly described, fantastically rare Vanessian, *Nymphalis danaus* Nab., orange-brown, with black-and-white foretips, mimicking, as its discoverer Professor Nabonidus of Babylon College, Nebraska, realized, not the Monarch butterfly directly, but the Monarch *through* the Viceroy, one of the Monarch's best known imitators.

This invented butterfly indicates no incestuous relationship nor interfamiliar breeding. It does, however, express an opinion.

The Monarch because of its absorption of noxious alkyloids in its larval food plant is inedible and is thus protected from many predators; the Viceroy through aeons of natural selection is the mimic, protected by its similarity in color and markings to the Monarch (Batesian mimicry). Presumably the *Nymphalis danaus* Nab. in those aeons would have evolved protection by mimicking *its* model, the Viceroy. This presupposes that the *N. danaus* did *not* habitate the same areas as the Monarch (for otherwise the Monarch would have been its direct model), and secondly, that it *did* habitate the same area as the Viceroy. Without the presence of the inedible Monarch, however, no predator would have hesitated to eat the edible nymphalids (Viceroy or *danaus*), and there would have been thus no purpose for mimicry. Natural selection in the evolution of the species would have been based on some other principle.

By naming the *danaus* after the danaid Monarch, its author would have recognized that it exceeded its model (the Viceroy) in simulating the Monarch, an excess in mimicry. The revealing *auctor* is attached to the name for complete scientific description. But the author, too, is bogus, mimicking the book's real author.

In this insect, Nabokov re-echoes his interest in taxonomy and voices an assessment of one part of a most important Darwinian theory, which relates to an aesthetic concept expressed in *Speak, Memory.*

> "Natural selection," in the Darwinian sense, could not explain the miraculous coincidence of imitative aspect and imitative behavior, nor could one appeal to the theory of the "struggle for life" when a protective device was carried to a point of mimetic subtlety, exuberance, and luxury far in excess of a predator's power of appreciation. I discovered in nature the nonutilitarian delights that I sought in art. Both were a form of magic, both were a game of intricate enchantment and deception. (p. 125)

Not only did Ada release (in one way or another) her larvae and pupae after Krolik's death in a symbolic gesture of turning from Lepidoptera, but also she was torn by dilemma, the conflict of her sexual life and the distortion of the relationship of art to natural science, and she abandons science for art (p. 193). Still, she maintains an ethereal link to her former interests. Entering the world of art through the theater, she assumes the stage name of Theresa Zegris, which, presumably, she has taken either from Chateaubriand or the Spanish Orange-tip butterfly, the Zegris. Van uses this epithet for her in his unmailed letter to her: "my Zegris," "my Spanish Orange-tip" (pp. 500-501).This Chateaubriand reference that pits the Abencerrajes family against the Zegris family is repeated in the "bogus" publishers of Van's book: the Zegris of London and the Abencerage of New York (p. 342). The Abencerraje butterfly (*Philotes abencerragus*: Lycaenidae) is totally unrelated to the Zegris.

If her theatrical life can be considered metaphorically analogous to the pupal stage, the wings developing and showing through the chrysalis, Ada is the one person in the novel who achieves complete metamorphosis. Her art and her natural marriage to Andrey Vinelander return her eventually to natural science. Once again she breeds and collects, now in a completely American setting. From Arizona she writes Van of the agave, "hosts of the larvae of the most noble animal in America, the Giant Skippers (Krolik, you see, is burrowing again)" (p. 385). This is a double allusion, as these larvae burrow at the base of their host food plant.

Still later, when she has become Mrs. Van Veen in a relationship where neither moral nor potential genetic problems exist, she integrates the larval imagery of the novel in her awareness that she is somehow responsible for "the metamorphosis of the lovely larvae that had woven the silk of Veen's Time" (p. 579). In a sense she has succeeded as creator herself, at least in literary conjunction. The patterns of the larvaless tree and the silken larva web on Van's lip (p. 95) have come full circle as literary device and in their lives.

Most important is Ada's metamorphosis in an aesthetic sense. Having become a photographer, she turned from preoccupation with larvae to filming butterflies.

> ... throughout her healthy and active old age [she] loved to film them in their natural surroundings, at the bottom of her garden, or at the end of the world, flapping and flitting, settling on flowers or filth, gliding over grass or granite, fighting, or mating. (pp. 567-68)

The reader is asked to remember that other "edge of the world" in her explorations of Van's genitals (p. 120). The reader should recognize, too, that Van's attitude is expressed in this passage; Ada would have been more specific, more poetic, and less degenerative. She has, in fact, by this time achieved true aesthetic appreciation of the beauty of mature butterflies, of art and natural history in general. Her note in Van's text corroborates her aesthetic eclosion.

> The song of a Tuscan Firecrest or a Sitka Kinglet in a cemetery cypress; a minty whiff of Summer Savory or Yerba Buena on a coastal slope; the dancing flitter of a Holly Blue or an Echo Azure—combined with other birds, flowers, and butterflies; *that* has to be heard, smelled and seen through the transparency of death and ardent beauty. (p. 71)

Holly Blue is the English name of *Celastrina argiolus,* related to the American Spring Azure (*C. ladon*) of which the Echo Azure is a subspecies. Blue butterflies in this case are part of an emotion that transcends mere happiness and approaches ecstasy.

Of all the characters in the book, Lucette is the most vulnerable, and she, too, is associated with a larval image. Van terms her "the impeccable paranymph" (p. 337). Her wearing of green and the "baby caterpillar" found on her clothes (p. 279) indicate that she is in the precarious larval stage, imprisoned by Van and Ada like the larvae in Ada's vivaria, and she is never to proceed beyond that stage in art or life. She is victimized, parasitized by Ada and Van—to their everlasting guilt. In remorse, Van transfers his imagery from butterflies to birds.

Van's larval associations, on the other hand, like those of his moths and butterflies, are an unresolved mixture of sensuality and irreality; his "esthetic empathy" is never fully realized. In retrospect only, is he moved by the larvae of Ada's vivaria: the "lovely, naked, shiny, gaudily spotted and streaked" caterpillars of the sharkmoth,[6] the flat catocala with its mimetic "grey knobs and lilac plaques" (the larva is true-to-life if the species is not authentic), the Vaporer "fellow" with its bright colors and "fancy tooth-brush tufts" (p. 55). The textures and colors forward his "texture of Time" theme and are also sexual. His *trompe l'oeil* tableaux (Forbidden Masterpieces) in the garden with Ada have a number of erotic images, some of which are oral. "There was a crescent eaten out of the vine leaf by a sphingid larva" (p. 141). He then recalls the "well-known" microlepidopterist who invented Marykisme, Adakisme, Ohkisme as species names for some of the multitudinous minute moths he was describing.

In that passage he questions the painting—swooning Satyr? Although he is not referring to satyrid butterflies, he had been told by Ada that those caterpillars that curl up upon being touched do not sleep but only "swoon" (p. 54). His most bizarre larva also swoons and curls up: "the six-inch long caterpillar, *qui rampait*" on a terrace overlooking a fabulous bay where Van is having his morning coffee with the Shah's pet dancer (p. 449). Van picks up this beautiful tufted creature and carries it to a bush, only to be paid for his trouble by having the "itchy bright hairs" stuck, cactus-like, in his fingers. The only recorded larvae of which the setae break off and remain in the skin is a species of Morpho, and Van was far from South America. Authentic fauna in Nabokov's writing do not move capriciously out of their geographic environments. That unglossy caterpillar has sexual innuendoes in his relationship to bacteria-like Eberthella Brown, but, more significantly, the incident illustrates that Van, even though he is a physician, is unable to deal adequately with all things that pertain to nature including both sex and art.

The majority of Van's Lepidoptera, larval or imaginal, are sexual in connotation.[7] In his first experience with Ada he likens her pubic hair to "a moth's shocking metamorphosis" (p. 59); this image is understandable only if one pictures the newly emerged moth, the wings crumpled, still wet with pupal fluid, incapable of flight until they are expanded by blood forced between the membranes and have dried—a very vulnerable insect. The next morning he ecstatically sees butterflies in everything, including "big bold Blues nearly the size of Small Whites and likewise of European origin" (p. 128). In putting these two butterflies in juxtaposition, Van has no real interest in either their comparative sizes nor their geographical origins. He is indicating that his present joy, represented by the blues (*iola*), carried over from the previous night, is not quite as great as the rapture of the sexual experience itself, represented by the pierid (*rapae*). That the Small White (known as the Cabbage butterfly in this country) has been successfully, and detrimentally, imported to America links the Euro-American scene of Antiterra through its fauna. The reader is told that forty years later, Van and Ada would again see the blues—in Switzerland, near Susten in the Valais. The bladder-senna (their larval food plant) butterflies of forty-years later are not shown to the readers, but their presence is made known in the name of Van's Villa Jolana, from the genus of these blues, *Iolana* (p. 552).

When Ada and Van are reunited at Mont Roux while Andrey is hospitalized, they see four different species of butterflies.

> The last butterflies of 1905, indolent Peacocks and Red Admirables, one Queen of Spain, and one Clouded Yellow, were making the most of the modest blossoms.   (p. 524)

Of these, three of the species are nymphalids and the last a pierid. Peacock Butterflies, considered by many to be the most handsome of Europe, are gregarious in the larval stage, weaving a communal web on the food plant (nettle) in which the caterpillars live and feed. The Red Admiral, as we have seen in *Pale Fire,* is both fatidic and emblematic, a Holarctic species known for its migratory habits. Its larval food plant is also nettle. Both colorful, even flashy, these two species, plural in the text, suggest the Zemski and Temnoisiniy families, which will end with Ada and Van.

The Queen of Spain, a fritillary of the kind that so interested Ada in her girlhood, suggests Ada herself. The name repeats the Spanish association that Van adopted from her performance in the Don-Quixote-Don-Juan movie.

The Clouded Yellow is also single, and following Nabokov's patterns, it may be presumed to be the pale yellow female with its greenish cast and black wing margins rather than the orange and black male of the species. The pale yellow butterfly would suggest the transcendent spirit of Lucette. The bird imagery, also associated with Lucette,[8] follows immediately in the text.

It is doubtful that Ada and Van saw the butterflies as other than part of the poetic landscape of their passion.

As Mason observes,[9] after Van knows that Ada has deceived him, moths appear more frequently in his imagery. At the family dinner (Marina, Demon, Ada, and Van) at Ardis in 1888, Van sees the moths as Ada's past and potential lovers, her "flutter-friends" that a "ghost" (Krolik) has pointed out. They are "pale intruders" with delicate wings; furry "ceiling-bumpers"; "thick-set rake-hells with bushy antennae"; and "party-crashing hawkmoths with red-black belted bellies" (p. 250). In personifying the moths, Van uses the pathetic fallacy for disparaging Ada's Lepidoptera interests as well as her lovers, and he correlates sex and tactile sensation.

Ada's and Van's sexual life is involved in the photograph that Kim took of the Peacock moths *in copula* (p. 400). There is a suggestion that Kim had also photographed Van and Ada, but attention is diverted by the mention of representations of copulating moths in actual paintings.

The pale, or white butterfly, as a symbol of approaching death is used by Van much as other Nabokovian characters experience the transcendent butterfly. Just before his duel, he sees one and knows "with utter certainty" that he has only moments to live (p. 310). But he does not die. Even his chosen symbol in his inverted patterns do not work as they do elsewhere in Nabokov's novels. Later, just before his reunion with Ada, he finds a "dead and dry hummingbird moth" (another sphingid) on the windowledge.[10] "Thank goodness, symbols did not exist either in dreams, or in life in between," he comments (p. 510). But he could have accepted the symbolism, for, indeed, Ada refuses to leave Andrey to join him.

"You have betrayed the Tree and the Moth! . . . *Oh! Qui me rendra mon Hélène...et le phalène*" (p. 530).

That he has seen symbols and they fail him evidence his wandering in and out of solipsism. Ultimately and out of time, he returns (inverts) to that position, the metaphor standing on its head, and will remain there at death. Van knew he was "not quite a savant but completely an artist" (p. 471). Unable to distinguish between artificial and real flowers, between the texture of Time and of larvae, he is an artist only in self-deception.[11]

The pale butterflies of the picnic chapter are not those of

transcendent spirits. The first, "a strange pale butterfly" (p. 79) appears just before the young siblings are to have their first bodily contact. Not specified, it denotes, like other of Nabokov's non-specific butterflies, love anticipated.

The second, "pale, diaphanous butterfly with a very black body" is a quasi-authentic species. Ada explains that it is "closely related to a Japanese Parnassian" (p. 85). She could simply have named the species; instead, she emphasizes a familial but certainly not a fraternal relationship. By pointing out the butterfly, she also draws the attention of the others away from her obviously emotional response to Van.

The Camberwell Beauty that Ada clutched (p. 170) at a point when Van is suspicious of a possible Lesbian relationship between Cordula and Ada is used in Nabokov's typical meaning with regard to the species. Ada and Van are soon to be separated for a long time.

Nabokov has used the larval-imaginal model as basic metaphor for *Ada,* and Van without consciously realizing it, has done the same for his "two" major books. He has given "new life to Time" by incorporating the larval image in his "kind of novella in the form of a treatise on the Texture of Time" (p. 562f). The larva is in essence the butterfly although it is not apparent in form. His family history, too, like the butterfly, transcends Time but nevertheless embodies the larva though there is no similarity to the *rampant* caterpillar.

"Our senses are simply not meant to perceive it. It is like—."

Ada has recognized through her intelligent understanding of nature and art the paradigm—and much, much more of Van's intentions.

"A Silent, Intense, Mimetic Pattern of Perfect Sense"[1]

In explaining the "Morpho-blue sky" metaphor used in *Lolita*, Nabokov said: " . . . I am thinking not of one of the many species of various blue Morpho butterflies of South America but of the ornaments made of bits of the showy wings of the common species."[2] Except for Paul Pilgram ("The Aurelian"), who would regret not being able to go to Brazil in search of these butterflies, and some references to color, the large blue butterflies (male, that is; the females are not predominantly blue), Nabokov uses the genus in ornaments. No doubt he would call them *poshlust,* as Ada speaks of the collage-pictures of disparate butterly wings as "always vulgar and often criminal" (*Ada,* p. 222)[3] and would include in this category the Urania moth mounted in plaster that Mademoiselle gave him as solace for his crushed collection (*Speak, Memory,* p. 128f). Van Veen refers to his new blue car as "dearer than sapphires and Morphos" (*Ada,* p. 551), even in middle-age still placing an artificial aesthetic value on nature. In *Look at the Harlequins!* Vadim remembers the glazed case with four morphos, and the satiny sheen is emphasized, having, like velvet, a sexual connotation.[4] The allusion is only partially to artificiality, as the case incarcerated the four morphos, like the convict that had captured them and is thus turned into a metaphor of imprisonment.

Unlike those morphos, Nabokov's blue butterflies, as we have seen, represent for the most part various degrees of ethereal happiness, whether they are the Karner Blues of Nabokov and Pnin, the Spring Azures of Ada, or the very special *L. sublivens* of Lolita. Not only was most of Nabokov's professional lepidopteran concern with these blue lycaenids, from the little *cormion* of France (described in his article and also in *Speak, Memory,*[5] p. 288) to the MCZ monograph of 1949, but also the very color was of particular importance in Nabokov's aesthetics. He used the physical principle in his biography of Gogol:

> As in the scaling of insects the wonderful color effect may be due
> not to pigment of the scales but to their position and refractive
> power, so Gogol's genius deals not in the intrinsic qualities of
> computable chemical matter (the "real life" of literary critics) but
> in the mimetic capacities of the physical phenomena produced by
> almost intangible particles of recreated life.[6]

The blue, as well as the white, and the iridescence on the scales of
butterfly wings is due to optical interference rather than pigmenta-
tion. The color is, in a sense, illusion to the eye while at the same time
being very real. Another illusionary trick of nature is apparent to man
and to Insectivora in the "eyespots" of many of the butterflies and
moths that Nabokov used in his writings: the Peacock Butterfly, the
Peacock moths, the Eyed Hawkmoth, the parnassians, and not least
the lycaenids.[7] The eyespots, found even on the larvae of some species,
such as the Tiger Swallowtail, are considered mimetic, startling and
diverting devices evolved for the protection of the species. Mimicry is
a partner in natural and in artistic deception, one of Nabokov's
repeated literary patterns.

As a lepidopterist Nabokov realized his dream of making a
scientific contribution by discovering and describing a new species.
His characters Krolik and Professor Nabonidus in *Ada* have been
pointed out as describers of species in more than one way. There are
others, such as the *Parnassius orpheus* Godunov, which Fyodor's
father described (*The Gift*, p. 124).[8] Actual species are the *apollo* and
the *phoebus*; the *orpheus* fits the mythology of Parnassus, and the life
of Godunov. He also described in a learned journal (1916) a most
peculiar creature: *Austautia simonoides,* a geometrid imitating a
parnassian (p. 115). The genus is named for Jean-Louis Austaut,
French lepidopterist, who described in 1889 variations, the *simonius*
and *simulator* of *Parnassius simo,* an Asiatic species. The terminal
ending-*noides* makes the butterfly Simo-like, and the invented species
is a take-off on the rather loose taxonomic practices of the nineteenth
century, practices that were eliminated by the International Code for
Scientific Nomenclature. Lepidopterist Paul Pilgram had a species
named for him: *Agrotis pilgrami* Rebel.[9] The generic name is valid,
the specific invented for the story. Actually there was a Viennese H.
Rebel who wrote on Lepidoptera with authority, but I suggest that
Nabokov has used this writer's name rather for disparaging reference

64

to another "rebel" doctor of Vienna, that is, Sigmund Freud.

In these and other of his fanciful creations, he found a sly way to make digs at theories and practices he considered equally spurious. For friendly spirits, however, he made up playful fancies, such as the *Papilio bunnyi* that he sketched and named for his friend Edmund Wilson[10] and the two that he drew for Alfred Appel, the *Flammea pallida* and the *Bonus bonus*.[11] His best invention, though, is the *Luperina berylae* Vandeleuri which he named in honor of Stapleton/ Vandeleur, the entomologist of *The Hound of the Baskervilles* (*Letters,* p. 162).

Taxonomy and systematics were, however, very serious subjects for him as were all his Lepidoptera activities. The "butterfly hunters" in his novels are self-parodies, and he speaks of humorous episodes in the activity in *Speak, Memory.* Even before his use of systematics as a theme in *Lolita,* he had used it in *The Eye.*[12]

> The situation was becoming a curious one. I could already count three versions of Smurov, while the original remained unknown. This occurs in scientific classification. Long ago, Linnaeus described a common species of butterfly, adding the laconic note *"in pratis Westmanniae."* Time passes, and in the laudable pursuit of accuracy, new investigators name the various southern and Alpine races of this common species, so that soon there is not a spot left in Europe where one finds the nominal race and not a local subspecies. Where is the type, the model, the original? Then, at last, a grave entomologist discusses in a detailed paper the whole complex of named races and accepts as the representative of the typical one the almost 200-year old, faded Scandinavian specimen collected by Linnaeus, and this identification sets everything right.   (p. 63)

This bit of the science of Lepidoptera classification on the narrator's part may seem irrelevant erudition; he was proud of that aspect of himself, as we are told on the last page. Later, however, the reference is explained. The narrator (Smurov) has been depicted by Roman Bogdanovich; it was a description that might emerge two-hundred years later, thus destined to live forever "to the delight of scholars" (p. 88). He has already described himself as having a blank conscience, "always exposed, always wide-eyed" (p. 17). This portrayal also fits the pale diaphanous Apollo parnassian with eye-like macula on its wings. He calls his image "a fetus in reverse" (p. 113), like a cocoon, and he has a definite preference for eau de cologne and for flowers. He sees all his associates as specimens that he can move about at will (i.e., pinned); they may exist only in his imagination, phantoms mirroring himself, like the multitude of races and subspecies described in that two-hundred-year interim.

The fact that time stops twice, once just before his suicide when he crushes his watch, and a second time in the dining room where Eugenia and Vanya have just found an unidentified object that might be made of wood (p. 70)—the hard, dark-brown cocoon of the Apollo Butterfly—reinforces the central theme.

The butterfly analogies of this novella could be carried further; indeed, the study could be carried to such extremes that the story could be reduced to a hunt for specification. In one sense it is just that, but the story is not allegory; the hunt for identification of the type (Smurov) and the paratypes is just one facet in the self-identification of the character of Smurov.

Nabokov parallels taxonomy to literary terms in his commentary on Pushkin's *Eugene Onegin*:

> As happens in zoological nomenclature when a string of obsolete, synonymous, or misapplied names keeps following the correct designation of a creature throughout the years and not only cannot be shaken off, or ignored, or obliterated within the brackets but actually grows on with time, so in literary history, the vague terms, "classicism," "sentimentalism," "romanticism," "realism," and the like straggle on and on, from textbook to textbook.[13]

That Nabokov could integrate his Lepidoptera interests into his concerns with literary arts and into his own art has been demonstrated. One point, however, his celebrated and repeatedly self-acknowledged lack of social-economic framework for his characters, Edmund Wilson found paradoxical.

Wilson asked Nabokov how he could justify his scientific interest in butterflies and their habitats and yet ignore the society and environment of the characters of his novels (*Letters*, p. 211). Nabokov replied that his interest was more on taxonomy, that "biological and ecological characters have no taxonomic value *per se*," noting that the same species of butterfly may breed in vastly different environments, separated by great distances, and that mankind is basically the same whether in China or in Egypt or "either of the two Georgias" (*Letters*, p. 214). He indicated this elsewhere, for instance in his introduction to

The Eye, explaining that his characters are Russian emigres in Berlin, London, or Paris but that they could be as well "Norwegians in Naples, or Ambracians in Ambridge." Certainly he recognized that the environments of revolution and displacement and other factors influence human action and thought; *Bend Sinister, Invitation to a Beheading,* even *Pnin* and *Pale Fire* recognize this, but more important to him as novelist than the individual's ecology are the universal pursuits of artistic expression and efforts to understand the absolutes that transcend space and time. As far as his own "influence of place" is concerned, he noted a debt to his native Russia only in "metaphors and sensuous associations."[14] Nationality itself is of secondary importance, as he explains through entomological analogy: "The more distinctive an insect's aspect, the less apt the taxonomist is to glance first of all at the locality label under the pinned specimen in order to decide which of several vaguely described races it should be assigned to."[15]

He eliminated a number of place names from his later editions of the English translation of *Camera Obscura* (*Laughter in the Dark*);[16] in the field, as evidenced by his reports of Lepidoptera trips, he took careful notes of habitat, food plants, locale and weather. In his articles he practiced population systematics, an aspect of the science which involves dimension of space and time. His Lepidoptera of the various literary genres are correct in season and place; even the fantasies of Antiterra are ecologically possible. He confesses that he does not believe in time (*Speak, Memory,* p. 139). In *Ada* he asks his readers to make the same transcendence over space and time that he used to illustrate the evolution of lycaenids.

By "straddling a Wellsian time machine" a modern taxonomist could explore the Cenozoic era and find in the Miocene some Asiatic butterflies still structually lycaenids but nothing distinguishable as Plebejinae, and yet "on a return journey" he would find structural differences that approximate generic diversity in the Palearctic butterflies of this family.[17]

In his own life, he erased the image of "memory's sting" as being that of the ichneumon wasp parasitizing its victim (*Speak, Memory,* p. 225), and translated time into memory, and memory into a butterfly, the *Parnassius mnemosyne* which illustrates the inside pages of his autobiography. Other Lepidoptera also spiral: the pugs of his childhood becoming the new find in Utah; those on the banks of the Oredezh transferring him to those of Colorado. None, however, can surpass the "intrepid butterfly" (as G. M. Hyde aptly terms it)[18] as the

Swallowtail that spirals from a childhood in Russia across the continents to Alaska and finally "to be captured after a forty-year race on an immigrant dandelion, under an endemic aspen near Boulder" (*Speak, Memory,* p. 120). To translate these life-long dedications, the science, art, and sport of Lepidoptera activity and the art and game of literature, into various genres not merely as trope but as intrinsic elements of his artistic vision is the ultimate spiral.

With the "inexorable laws" of systematics and the constant attempt to define species and subspecies in meaningful terms, most of the species he described are now classed as subspecies. In popular nomenclature none bears his name because he had described it. Lepidopterists, though, have long memories, too, and have chosen to honor him by designating species in his name: Nabokov's Pug (James H. McDunnough, 1945), Nabokov's Blue (F. Martin Brown, 1955); Nabokov's Satyr (Robert M. Pyle, 1981).

Postscript

As a final word to this study, I add a scientist's advice, which has, I believe, application not only to biological observation but also to literary analysis.

> When closely—no matter how closely—observing events in nature, we must, in the very process of observation, beware of letting our reason—that garrulous dragoman who always runs ahead—prompt us with explanations which then begin imperceptively to influence the very course of observation and distort it; thus the shadow of the instrument falls upon the truth.
>
> K. K. Godunov-Cherdyntsev.
> Vladimir Nabokov, *The Gift,* p. 342

Notes

Chapter 1

1. The exchange with Alvin Toffler that appeared in *Playboy* (January 1964) is reprinted in *Strong Opinions* (New York: McGraw-Hill, 1973), pp. 20-45.

2. Vladimir Nabokov, *Speak, Memory, an Autobiography Revisited*. Rev. ed. (New York: G. P. Putnams, 1966). Textual references are to this edition.

3. Vladimir Nabokov, *The Gift* (New York: Capricorn Books, 1970; reprint of 1963 edition). Textual references are to this edition.

4. Vladimir Nabokov, *Look at the Harlequins!* (New York: McGraw-Hill, 1974). Textual references are to this edition.

5. Vladimir Nabokov, *Ada* (New York: McGraw-Hill, 1969).

6. Vladimir Nabokov, *King, Queen, Knave* (New York: McGraw-Hill, 1968). Textual references are to this edition.

7. Herbert Grabes, Dabney Stuart, and a number of others have pointed out the fictive qualities of this autobiography.

8. In an interview with Herbert Gold published in *The Paris Review* (October 1967) and reprinted in *Strong Opinions,* p. 100.

9. Nikolai I. Kuznetsov (1873-1948), Russian entomologist, published his *Faune de la Russie et des Pays Limitrophes,* Vol. 1, *Insectes Lepidoptères* in 1915 (mentioned in *Speak, Memory,* p. 133). Preceding him in publications on the Lepidoptera of Russia was G. Fischer von Waldheim, who is referred to in *The Gift,* p. 114.

10. Vladimir Nabokov, "A Few Notes on Crimean Lepidoptera," *The Entomologist,* 53 (no. 680, January 1920), pp. 29-33.

11. Vladimir Nabokov, "Notes on the Lepidoptera of the Pyrenées Orientales and the Ariège," *The Entomologist,* 64 (1931), pp. 255-57, 268-71.

12. "Some people intent on the process of digestion kept strolling up and down in the darkness with glowing cigars." Ibid., p. 255.

13. Vladimir Nabokov, "The Aurelian," in *Nabokov's Dozen* (Freeport: Books for Libraries Press, 1969; reprint of Doubleday, 1958), pp. 95-112. The term "aurelian" stems from the gold of certain butterfly chrysalids and evolved into a synonym for lepidopterist.

14. Vladimir Nabokov, "Christmas," in *Details of a Sunset and Other Stories* (New York: McGraw-Hill, 1976), pp. 151-62.

15. William W. Rowe, *Nabokov's Spectral Dimension* (Ann Arbor: Ardis, 1981) analyzes Nabokov's spirits, acting through Lepidoptera.

16. Vladimir Nabokov, *Poems and Problems* (New York: McGraw-Hill, 1970), pp. 46-47.

17. Vladimir Nabokov, "*Lysandra cormion,* a New European Butterfly," *Journal of the New York Entomological Society,* 49 (1941), pp. 265-67.

18. Vladimir Nabokov, *Pnin* (New York: Doubleday, 1957). Textual references are to this

edition.

19. Originally published in *The New Yorker* in 1942 and as "A Discovery" in *Poems* (New York: Putnams, 1959), p. 15, and in *Poems and Problems* (pp. 155-56).

20. Vladimir Nabokov, "Some New or Little Known Nearctic *Neonympha* (Lycaenidae: Lepidoptera)." *Psyche,* 50 (1945), pp. 61-80.

21. Vladimir Nabokov, "Notes on Neotropical Plebejinae (Lycaenidae: Lepidoptera). *Psyche,* 50 (1945), p. 4f.

22. Ibid., pp. 1-61.

23. Vladimir Nabokov, *The Nabokov-Wilson Letters, 1940-1971,* edited by Simon Karlinsky (New York: Harper, 1979), p. 240.

24. Vladimir Nabokov, "The Nearctic Members of the Genus *Lycaeides* Hübner." *Bulletin of the Harvard Museum of Comparative Zoology,* 101, no. 4 (1949), pp. 249-541.

25. *The Nabokov-Wilson Letters,* p. 131.

26. Ibid., p. 116.

27. Vladimir Nabokov, *Bend Sinister* (New York: McGraw-Hill, 1974; reprint of New Directions, 1947). Introduction, p. v. Textual references are also to this edition.

28. *The Nabokov-Wilson Letters,* p. 127.

29. Ibid., p. 106. In the text I cite there is a mistake in the spelling of the genus, which I have corrected here. Nabokov discovered Nabokov's Pug at James Laughlin's lodge in Utah in 1943. (*Speak, Memory,* pp. 127-28).

30. In the interview with Alfred Appel, Jr. published in *Wisconsin Studies in Contemporary Literature,* Spring 1967 and reprinted in *Strong Opinions,* p. 78f.

31. In the interview with Kurt Hoffman of October 1971 and printed in *Strong Opinions,* p. 190.

32. Interview with Philip Oates, *The Sunday Times,* London (June 22, 1969) and reprinted in *Strong Opinions,* p. 136.

Chapter 2

1. Jane Grayson, *Nabokov Translated* (Oxford: Oxford University Press, 1977), p. 146.

2. Vladimir Nabokov, *Mary* (New York: McGraw-Hill, 1970). Textual references are to this edition.

3. Vladimir Nabokov, *Glory* (New York: McGraw-Hill, 1971). Textual references are to this edition.

4. Vladimir Nabokov, *King, Queen, Knave* (New York: McGraw-Hill, 1968). Textual references are to this edition.

5. Vladimir Nabokov, *Speak, Memory,* Rev. ed. (New York: Putnams, 1966). Textual references are to this edition.

6. Vladimir Nabokov, "Foreword" to *King, Queen, Knave,* p. viii.

7. William W. Rowe, *Nabokov's Spectral Dimension* (Ann Arbor: Ardis, 1981) sees these somewhat differently. P. 113.

8. G. M. Hyde, *Vladimir Nabokov, America's Russian Novelist* (London: Marion Boyars, 1977), p. 60 *passim.* The black-white images of *Laughter in the Dark* are used similarly here.

9. Vladimir Nabokov, *Despair* (New York: McGraw-Hill, 1965). Textual references are to this edition.

10. Vladimir Nabokov, *The Defense* (New York: Capricorn Books, Putnams, 1970). Textual references are to this edition.

11. There may be a good pun on thyme-time, but these butterflies feed on the nectar of the blossoms, and thyme is the larval food plant of several European lycaenids.

12. Grayson, op. cit., p. 34.

13. Vladimir Nabokov, *Transparent Things* (New York: McGraw-Hill, 1972). Textual references are to this edition.

14. Rowe, op. cit., p. 15.

15. Vladimir Nabokov, *Look at the Harlequins!* (New York: McGraw-Hill, 1974). Textual references are to this edition.

16. Herbert Grabes, *Fictitious Biographies: Vladimir Nabokov's English Novels* (The Hague: Mouton, 1977), makes this observation in his interpretation of *"metamorphoza."* P. 124.

Chapter 3

1. Vladimir Nabokov, *Invitation to a Beheading* (New York: Capricorn Books, 1965. Reprint of 1959), p. 7. Textual references are to this edition.

2. Vladimir Nabokov, *Bend Sinister* (New York: McGraw-Hill, 1974. Reprint of New Directions, 1947). Textual references are to this edition.

3. Vladimir Nabokov, Foreword to *Invitation to a Beheading*, p. 7.

4. Vladimir Nabokov, *Speak, Memory,* Revised edition (New York: Putnams, 1966). Textual references are to this edition.

5. Vladimir Nabokov, *The Gift* (New York: Capricorn Books, 1970). Textual references are to this edition.

6. William W. Rowe, *Nabokov's Deceptive World* (New York: New York University Press, 1971), pp. 181-86.

7. Vladimir Nabokov, *The Real Life of Sebastian Knight* (Norfolk, Conn.: New Directions, 1941, 1959). Textual references are to this edition.

8. Vladimir Nabokov, *Ada* (New York: McGraw-Hill, 1969).

9. Page Stegner, *Escape into Aesthetics, the Art of Vladimir Nabokov* (New York: Dial Press, 1966), p. 67. Stegner was, I believe, the first to point out this reference.

10. Vladimir Nabokov, *Pnin* (Garden City: N.Y.: Doubleday, 1957). Textual references are to this edition.

11. Julia Bader, *Crystal Land, Artfice in Nabokov's English Novels* (Berkeley: University of California Press, 1972), p. 82 *passim*.

12. William Carroll, "Nabokov's Signs and Symbols," in *A Book of Things about Vladimir Nabokov,* edited by Carl Proffer (Ann Arbor: Ardis, 1974), p. 206.

Chapter 4

1. Vladimir Nabokov, *The Gift* (New York: Capricorn Books, 1970. Reprint of 1963). Textual references are to this edition.

2. Vladimir Nabokov, *Speak, Memory.* Revised edition (New York: Putnams, 1966). Textual references are to this edition.

3. Carol Williams, "Nabokov's Dialectical Structure," in *Nabokov, the Man and His Work,* edited by L. S. Dembo (Madison: University of Wisconsin Press, 1967) explains that in the rainbow's arc, according to Nabokov, the human eye sees only half the circle and the other half must be taken on faith (p. 165).

4. Jane Grayson, *Nabokov Translated* (Oxford: Oxford University Press, 1977) shows that in the Russian original Nabokov used "silkworm" which would give it an entirely different meaning and add an oriental image.

5. G. M. Hyde, *Vladimir Nabokov, America's Russian Novelist* (London: Marion Boyars, 1977), pp. 21-22.

6. J. W. Tutt was not the discoverer of the sphragus of the parnassian. Active in taxonomy, he is known for his work, *A Natural History of British Lepidoptera* (1899).

7. The model for this missionary may be Father Armand David, French naturalist. See Simon Karlinsky's note 8 to Letter 190 in *The Nabokov-Wilson Letters, 1940-1971* (New York: Harper, 1980), p. 265.

Chapter 5

1. Vladimir Nabokov, *Pale Fire* (New York: Putnams, 1962). Textual references are to this edition.

2. Julia Bader, *Crystal Land, Artifice in Nabokov's English Novels* (Berkeley: University of California Press, 1972), p. 45.

3. James G. Frazer, *The Golden Bough*. Edited by Theodore H. Gaster (New York: Criterion Books, 1959), p. 309ff.

4. Nonetheless, the French vernacular name for the species is *Le Flambé*.

5. While butterflies commonly frequent flowers for the nectar, many do alight on rotting fruit, and flesh, and dung. Thomas C. Emmel writes: "These unusual sources appear to give the butterflies amino acids—the building blocks of protein—and other essential nitrogen-containing substances." *Butterflies: Their World, Their Life Cycle, Their Behavior* (New York: Knopf, 1975), p. 55.

6. Vladimir Nabokov, in an interview with Alfred Appel, Jr., August 1970, reprinted in *Strong Opinions* (New York: McGraw-Hill, 1973), pp. 169-70.

Chapter 6

1. Vladimir Nabokov, *Lolita* (New York: Putnams, 1955). Textual references are to this edition.

2. Alfred Appel, Jr., Introduction to *The Annotated Lolita* by Vladimir Nabokov. Edited by Alfred Appel Jr. (New York: McGraw-Hill, 1970), p. lix.

3. Diana Butler, "Lolita Lepidoptera," *New World Writing*, 16 (Philadelphia: Lippincott, 1960), pp. 58-84.

4. Alfred Appel makes this observation, *The Annotated Lolita*, Notes, p. 360.

5. William Rowe, *Nabokov's Deceptive World* (New York: New York University Press, 1971) interprets this as a sexual play on words, p. 130.

6. Appel, op. cit., p. lx.

7. Ibid., p. 376.

8. Ibid., p. 383.

9. Ibid., p. 384.

10. Vladimir Nabokov, "The Female of *Lycaeides sublivens* Nab.," *The Lepidopterists' News*, 6 (1952), p. 36.

11. *The Annotated Lolita*, p. 408.

12. Ibid., p. 393.

13. Ibid., p. 409.

14. Ibid., p. 422.

15. William W. Rowe, *Nabokov's Spectral Dimension* (Ann Arbor: Ardis, 1981) finds spirits acting through butterflies and moths.

16. Compare: Ernst Mayr, *Systematics and the Origin of Species* (New York: Columbia University Press, 1942), p. 3: "There was a tendency /1900-1930/ among laboratory workers to think rather contemptuously of the museum man who spent his time counting hairs or drawing bristles and whose final aim seemed to be merely the correct naming of the species."

Chapter 7

1. Vladimir Nabokov, *Ada or Ardor: a Family Chronicle* (New York: McGraw-Hill, 1969). Textual references are to this edition.

2. Not only are pupae and puppy assonantly related, but while the English "caterpillar" has a feline association, the French *chenille* has a canine etymology (from Latin *canicula*). I believe

Nabokov is making this allusion in his combination of animals here and in *Bend Sinister*, p. 27.

3. Bobbie Ann Mason, *Nabokov's Garden, a Guide to Ada* (Ann Arbor: Ardis, 1974): "Ada's larvarium and her ambitions for breeding are expressions of her natural desire for procreation, a predilection which finds an analogy in the orchids she fancifully crossbreeds," p. 56.

4. An interesting discussion of the names of the catocalas is given by Theodore D. Sargent in his *Legions of Night: The Underwing Moths* (Amherst: University of Massachusetts Press, 1976), pp. 1-11. Dr. Sargent, by the way, proposes naming a melanic form of *C. micronympha* the "lolita" (p. 74).

5. Mason explains the fertility of Krolik's name, op. cit., p. 56.

6. Compare *Speak, Memory*: "the discreet, pleasantly cool, rhythmically undulating caress of a caterpillar ascending one's bare skin" (New York: Putnams, 1966), p. 196.

7. "The butterfly motif expands with the children's awakening love." Mason, op. cit., p. 61.

8. Ibid., p. 64 *passim.*

9. The Lepidoptera "undergo a relegation in status and moths appear more frequently." loc. cit.

10. The hummingbird moths are not only sphingids and therefore in Nabokov's novels carry the metaphor of sexual passion but also this insect unites the lepidopteran and ornithological imagery of the novel, especially in the hummingbirds and sunbirds associated with Ada.

11. Mason, op. cit., pp. 105-7.

Chapter 8

1. Vladimir Nabokov, "The Poem." Published in *Poems* (Garden City, N.Y.: Doubleday, 1957), p. 17 and reprinted in *Poems and Problems* (New York: McGraw-Hill, 1970), p. 157.

2. Quoted in notes to *The Annotated Lolita,* edited by Alfred Appel, Jr. (New York: McGraw-Hill, 1970), p. 405f.

3. Vladimir Nabokov, *Ada* (New York: McGraw-Hill, 1969).

4. William W. Rowe, *Nabokov's Deceptive World* (New York: New York University Press, 1971), p. 111 suggests that the influence of the Russian word which conveys the meaning of "flow" and "pouring off" lingers in Nabokov's use of "sheen."

5. Vladimir Nabokov, *Speak, Memory.* Revised ed. (New York: Putnams, 1966). Textual references are to this edition.

6. Vladimir Nabokov, *Nikolai Gogol* (Norfolk, Conn.: New Directions, 1944), p. 56.

7. The defraction in the eyespot on a butterfly's wing is used in literary theory by Nabokov in his essay "The Art of Literature and Commonsense," in *Lectures on Literature,* edited by Friedson Bowers (New York: Harcourt, Brace, Jovanovich, 1980), p. 374f.

8. Vladimir Nabokov, *The Gift* (New York: Capricorn Books, 1970; reprint of 1963). Textual references are to this edition.

9. Vladimir Nabokov, "The Aurelian," in *Nabokov's Dozen* (Freeport, N.Y.: Books for Libraries Press, 1969; reprint of 1958).

10. Vladimir Nabokov and Edmund Wilson, *The Nabokov-Wilson Letters, 1940-1970,* edited by Simon Karlinsky (New York: Harper, 1980), p. 122. Textual references are to this edition.

11. Alfred Appel, Jr., "Backgrounds of Lolita," in *Nabokov; Criticism, Reminiscences, Translations, and Tributes,* edited by Alfred Appel, Jr. and Charles Newman (Evanston: Northwestern University Press, 1970), p. 28. The sketches were reprinted in *Time,* May 23, 1969.

12. Vladimir Nabokov, *The Eye* (New York: Phaedra Press, 1965). Textual references are to this edition.

13. Alexander Pushkin, *Eugene Onegin, a Novel in Verse.* Translated with a commentay by Vladimir Nabokov (New York: Bollingen Foundation, Pantheon Press, 1964), Volume III, p. 32.

14. Vladimir Nabokov, interview with Jane Howard, published in *Life* (November 20, 1964), reprinted in *Strong Opinions* (New York: McGraw-Hill, 1976), p. 46.

15. Vladimir Nabokov, interview with Alfred Appel, Jr., reprinted in *Strong Opinions*, p. 63.

16. Grayson, op. cit., p. 33.

17. Vladimir Nabokov, "Notes on Neotropical Plebijinae," *Psyche,* vol. 52 (1945), p. 44.

18. G. M. Hyde, *Vladimir Nabokov, America's Russian Novelist* (London: Marion Boyars, 1977), p. 195.

Appendix A

Nabokov's Lepidoptera: Species

This is intended as a guide to the species that Nabokov uses in his literary works, including the interviews in *Strong Opinions,* and in his letters to Edmund Wilson. Excluded are the species in his articles on Lepidoptera, his strictly scientific works (except as they are alluded to in the various genres, letters, and interviews), and his book reviews, even those reprinted in *Strong Opinions.*

The entry is listed as Nabokov names the species, whether by scientific name or by vernacular name. In those cases where the insect is described but not named, it is listed by its scientific name in brackets.

A selected bibliography of useful field guides and reference works on butterflies and moths is also appended in hopes that the reader will view the photographs and plates of Nabokov's species and perhaps pursue further the study of Lepidoptera.

/*Agrius convolvuli*/ (Sphingidae). Convolvulus Hawkmoth. Widespread across Europe to Africa, Asia, Australia; absent from the Americas. *Ada,* pp. 56-57.

/*Aletis libyssa*/ (Geometridae). The African moth that mimics *Danaus chrysippus* in a complex of mimicry that involves several other moths and several genera of butterflies. *The Gift,* p. 122.

Amandus Blue. *Plebicula amanda* (Lycaenidae). (Amanda's Blue). A European Blue, ranging into North Africa. I cannot account for Nabokov's changing the gender of the vernacular; he was fastidious in preserving the scientifc gender, *viz.* his reference to *Hemiargus isola* in his article on Wyoming butterflies. *The Gift,* p. 145.

Amur Hawkmoth. *Smerinthus tremulae amurensis* (Sphingidae). Eastern Europe into China along the Amur River. A handsome, grey-marbled moth, "which in happier days, I used to find at the foot of aspens in the neighborhood of Petrograd" ("A Few Notes on Crimea Lepidoptera," *Ent.,* 1920). *Speak, Memory,* p. 156.

Anglewing. See Comma Butterfly.

Aphantopus Ringlets. *Aphantopus hyperantus* (Satyridae). Europe into Asia. *The Gift,* p. 144.

Apollo Butterfly. *Parnassius apollo* (Papilionidae). Mountainous areas of continental Europe. "The Aurelian," in *Nabokov's Dozen,* p. 104; "Slava" ("Fame") in *Poems and Problems,* p. 105; possibly that of *Transparent Things,* p. 90. (Other Apollos are the Black and the Imperatorial.)

Aricia psilorita (Lycaenidae). The Cretan Argus. Known only from Mt. Ida, the highest point on Crete. Nabokov referred to the labels of his series of this species to determine precise altitude. *The Nabokov-Wilson Letters,* p. 158.

Arran Browns. *Erebia ligea* (Satyridae). Europe into Asia. *Speak, Memory,* p. 131.

Aspen Hawk Moth. *Laothoë populi* (Sphingidae). The Poplar Hawkmoth. Palearctic. Larval foodplants are aspen and poplar (*Populus*); habitat is moist woodland, river banks. *The Gift*, p. 121.

Atlantis. *Speyeria atlantis* (Nymphalidae). The Atlantis Fritillary. Wide range in North America. *Pale Fire*, p. 169.

Attacus Moth. *Attacus atlas* (Saturnidae). Called also the Atlas Moth, the Giant Atlas Silkmoth. Indo-Australian Region. "Christmas," *Details of a Sunset*, p. 161.

Bird-wing Butterflies. *Ornithoptera* (genus) (Papilionidae). Large butterflies of the Indo-Australian region. Highly prized by collectors. *Transparent Things*, p. 4.

Black Apollo. *Parnassius mnemosyne*, q.v.

Black Ringlet. *Erebia melas* (Satyridae). Southeastern Europe. *The Gift*, p. 121.

Black-Veined Whites. *Aporia crataegi* (Pieridae). Continental Europe, North Africa, Asia. (Extinct in England). *The Gift*, p. 144.

Brimstone. *Gonepteryx rhamni* (Pieridae). A common sulphur butterfly in Europe; also North Africa, Asia. *The Gift*, p. 36; *Speak, Memory*, p. 111. (See also: Cleopatra Brimstone).

Burnet Moth. (Zygaenidae). There are a number of blue and red species with blue antennae among this family of European small, day-flying moths. *The Gift*, p. 144.

Butler's Pierid. *Baltia butleri* (Pieridae). Tibet. *The Gift*, p. 133.

Cabbage Butterfly. *Artogeia rapae* (Pieridae). Known in Europe as the Small White. Introduced into North America. *Mary*, p. 91; *The Gift*, p. 145; *Ada*, p. 128.

Callophrys sheridanii Edwards. (Lycaenidae). The White-Lined Green Hairstreak. Western North America from British Columbia to Saskatchewan, south to New Mexico and Arizona. *The Nabokov-Wilson Letters*, p. 106.

Camberwell Beauty. *Nymphalis antiopa* (Nymphalidae). Holarctic. Known in North America as the Mourning Cloak. "Christmas" *Details of a Sunset*, p. 159; *Mary, p. 60; The Real Life of Sebastian Knight*, p. 139; *Speak, Memory*, p. 231, p. 239; *Ada*, p. 170.

Catocala adultera (Noctuidae). Russia. The underwings are red, similar to *C. nupta. Speak, Memory*, p. 135.

/*Catocala fraxini*/ (Noctuidae). The Blue Underwing; the Clifden Non-pareil. Europe. *The Gift*, p. 107.

Cercyonis behrii Grinnell = *Cercyonis sthenele sthenele* (Satyridae). Subspecies from San Francisco, now extinct because of habitat destruction. *The Nabokov-Wilson Letters*, p. 106.

Chapman's Hairstreak. *Callophrys avis* (Lycaenidae). Chapman's Green Hairstreak. Southwest Europe and North America. *Speak, Memory*, p. 205.

/*Chrysiridia ripheus*/ (Uraniidae). Day-flying moth with colorful irridescent wings and tails. Madagascar. "The Aurelian," in *Nabokov's Dozen*, p. 106.

Clearwing Moths. (Sesiidae; syn. Aegeriidae). Diurnal moths that mimic wasps or bees. Most scales are lost in first flight, leaving the wings transparent. Worldwide. "The Aurelian" in *Nabokov's Dozen*, p. 151.

Cleopatra. *Gonepteryx cleopatra* (Pieridae). A brimstone of southern Europe, North Africa to Syria. *Speak, Memory*, p. 147.

Clifden Blue. *Lysandra bellargus* (Lycaenidae) = the Adonis Blue. Named the Clifden Blue by Moses Harris in *The Aurelian* (1766) from the locality in which it was found (Cliveden, Buckinghamshire, England). *Strong Opinions*, p. 60.

Clouded Yellow. *Colias crocea* (Pieridae). Europe, parts of Asia. *Ada*, p. 524; *Speak, Memory*, p. 147.

Comma Butterfly. *Polygonia c-album* (Nymphalidae). Throughout Europe into Asia. The North American Comma Butterfly is *P. comma. The Gift*, p. 344; *Speak, Memory*, p. 166.

Cordigera. *Anarta cordigera* (Noctuidae). Common Yellow Underwing. Also called the Catocaline Anarta by W. J. Holland. Holarctic. *Speak, Memory*, p. 138.

Corsican Swallowtail. *Papilio hospiton* (Papilionidae). Corsica, Sardinia. "The Aurelian" in *Nabokov's Dozen*, p. 104f.

Cowl Moth. See Sharkmoth.

Cyclargus arembis Nabokov = *Hemiargus thomasi arembis* (Lycaenidae). Subspecies of the Miami Blue. Nabokov described it from a specimen from the Cayman Islands, West Indies. *The Nabokov-Wilson Letters,* p. 169.

Death's Head Hawk Moth. *Acherontia atropos* (Sphingidae). So called because of the appearance of a human skull in the pattern of the scales on its thorax. Europe. *The Gift,* p. 122.

Diana. *Speyeria diana* (Nymphalidae). A fritillary butterfly of eastern United States. *Pale Fire,* p. 169.

Echo Azure. *Celastrina ladon echo* (Lycaenidae). California subspecies of the Spring Azure. *Ada,* p. 71.

Egerias. See *Pararge aegeria*

Elwes Swallowtail. *Papilio elwesi* (Papilionidae). A "black wonder with tails in the shape of hooves." China, Formosa. *The Gift,* p. 134.

Emperor Moth. *Saturnia pavonia* (Saturnidae). Nabokov explains the confusion between this moth and the Purple Emperor Butterfly (q.v.). Europe into Asia. *The Nabokov-Wilson Letters,* p. 113.

Epicnaptera arborea. (Lasiocampidae). A lappet moth. Godunov brought back a specimen from Siberia only to find them also in St. Petersburg. *The Gift,* p. 107.

Ergane. *Pieris ergane* (Pieridae). The Mountain Small White. Southern Europe into Asia Minor. *Look at the Harlequins!,* p. 35f.

/*Euchloe ausonia*/ (Pieridae). The Mountain Dappled White. The rich amateur Sommer was so knowledgeable that he would identify the subspecies *uralensis,* which Nabokov found in the Crimea. "The Aurelian," *Nabokov's Dozen,* p. 106.

Eupithecia nabokovi McDunnough (Geometridae). Nabokov's Pug. A small moth of western United States, which he collected in Utah and which was named in his honor. *Speak, Memory,* p. 126.

Everes comyntas (Lycaenidae). The Eastern Tailed Blue. Common from southern Canada into the Rockies. *The Nabokov-Wilson Letters,* p. 101.

Freya Fritillary. *Clossiana freija* (Nymphalideae). Northern Europe into Japan; also North America. The "dusky little fritillary" with the name of a Norse goddess. *The Gift,* p. 145; *Speak, Memory,* p. 138.

Giant Skippers. Family Megathymidae of the American Southwest. Agave is the larval host plant of many species. *Ada,* p. 385.

Goat Moth. *Cossus cossus* (Cossidae). The caterpillars have a goat-like odor. Europe, North Africa, central Asai. *Speak, Memory,* p. 132.

Green-veined Whites. *Artogeia napi* (Pieridae). North Amrica, Europe, Asia to Japan. The American "Mustard White," or Veined White. *Strong Opinions,* p. 60.

Grinnell's Blue. *Glaucopsyche lygdamus australis* Grinnell (Lycaenidae). A subspecies of the Silvery Blue. Southern California. *Strong Opinions,* p. 182.

Grüner's Orange-Tip. *Anthocharis grueneri* (Pieridae). Southern Europe. *Speak, Memory,* p. 253.

Hairstreak. See *Strymonidia W-album*

Hawkmoth (Hawk Moth). A common designation for some Sphingidae. See Oleander Hawk; Death's Head Hawk Moth; *Smerinthus ocellata; Agrius convulvuli,* and Hummingbird Moth.

Heldreich's Sulphur. *Colias aurorina heldreichi* (Pieridae). Subspecies of the Dawn Clouded Yellow. Only on Mt. Veluchi, Greece. *Speak, Memory,* p. 253.

Hero Ringlet. See Scarce Heath

/*Hesperis comma*/ (Hesperidae). Known as the Silver-Spotted Skipper in England, as the Comma Skipper in North America, it has many forms and subspecies. *The Gift,* p. 346.

Hippolyte Grayling. *Pseudochazara hippolyte* (Satyridae). Identified as the Euxine race in the text. Also called the Nevada Grayling because of its population in the Sierra Nevada of

Spain; disjunct populations also in southern Russia. Also Asia Minor, east to China. *Speak, Memory*, p. 247.

Holly Blue. *Celastrina argiolus* (Lycaenidae). Palearctic.

Hummingbird Moth. Also called Hummingbird Hawk Moth. Some authorities make a distinction between those Sphingidae with rather longer probosces, useful for probing into tubular flowers, a characteristic they share with hummingbirds, and the Hawk Moths of the same family. Other authorities consider the terms synonomous. Like many hawks and to some extent hummingbirds, these moths hover. Among the various genera of Sphingidae, there are those with clear wings and those with densely scaled, opaque wings. The "glasslike" wings of the species in *The Gift*, p. 145, leads to identification of *Hemaris fuciformis*, a common European species; that of *Speak, Memory*, p. 134 may be *Proserpinus proserpina* (the Willowherb Hawk Moth). The one in *Ada*, p. 510, is not described; the "grey hummingbirds" of *Lolita*, p. 159 *Hyles lineata*.

Imperatorial Apollo. *Parnassium imperator* (Papilionidae). China, Tibet. *The Gift*, p. 135.

/*Iolana iolas*/ (Lycaenidae). The Iolas Blue. These "big bold Blues" are widespread in Europe and North Africa. Larval food plant is bladder-senna (*Colutea aborescens*) *Ada*, p. 128.

/*Iphiclides podalirius*/ (Papilionidae). The Scarce Swallowtail. Central and southern Europe, North Africa, temperate Asia. *Glory*, p. 22.

Karner Blue. See *Lycaeides melissa samuelis* Nabokov

Krueper's White. *Pieris krueperi* (Pieridae). Greece and Asia Minor. *Speak, Memory*, p. 253.

Large Copper. *Lycaena dispar dispar* (Lycaenidae). R. M. Pyle notes the status of this butterfly as follows: "The English, nominate subspecies *L. dispar dispar* became extinct in the 19th century due to drainage of the fens. Now one small colony of introduced German *L. dispar batavus* is maintained on a Fen nature reserve. The species as a whole is endangered on the continent due to wetland drainage." (Quoted with permission). *The Real Life of Sebastian Knight*, p. 43.

Large Emerald. *Geometra papilionaria* (Geometridae). A large green moth common throughout Europe. *Speak, Memory*, p. 132.

Large White. *Pieris brassicae* (Pieridae). North Africa, western Europe, across Asia. *Speak, Memory*, p. 127.

Libytheana bachmanii (Libytheidae). The Snout Butterfly. A "very rare migrant" Nabokov wrote from Ithaca, N.Y. A North American species, ranging into Mexico, the species is resident in the South and strays into the Midwest and Northeast. It is easily identifiable at some distance. *The Nabokov-Wilson Letters*, p. 205.

Limenitis populi (Nymphaliedae). The Poplar Admiral. Larval food plant is poplar, aspen. Widespread but uncommon in Europe (not in British Isles) and Asia. Nabokov also mentions *L. populi bucovinensis* Hormuzaki, the Russian Poplar Admiral. *The Gift*, p. 90; *Speak, Memory*, p. 135, 192.

Lobster Moth. *Stauropus fagi* (Notodontidae). Europe, Asia to Japan. Its vernacular name refers to the "acrobatic caterpillar," which somewhat resembles a lobster. *Speak, Memory*, p. 124.

Luna Moth. *Actias luna* (Saturnidae). Also called the Moon Moth. Eastern United States, Mexico. One of its larval food plants is hickory. *Pale Fire*, p. 114.

Lycaeides Hübner. The genus of Lycaenidae in which Nabokov made his most significant scientific contributions in two articles and one monograph. *The Nabokov-Wilson Letters*, pp. 115, 206, 210.

Lycaeides anna Edwards = *Lycaeides argyrognomon anna* Edwards (Lycaenidae). Nabokov reports that he has proved that this is a subspecies in the *argyrognomon* group rather than the *melissa* group. Northern California, southern Oregon, western Nevada. *The Nabokov-Wilson Letters*, p. 94.

Lycaeides argyrognomon Bergstrasser (Lycaenidae). The Northern Blue in North America; Reverdin's Blue in Continental Europe. There are several subspecies. See *L. Anna; L.*

80

sublivens; L. scudderi lotis. The Nabokov-Wilson Letters, p. 94.

Lycaeides cleobis (Lycaenidae). Small blue from Siberia to North Korea. Nabokov postulates the existence of a subspecies in North America, but this is not substantiated. *The Nabokov-Wilson Letters,* p. 94.

Lycaeides melissa Edwards (Lycaenidae). The Melissa Blue; the Orange-Bordered Blue. Wide range through central North America. *The Nabokov-Wilson Letters,* p. 94.

Lycaeides melissa samuelis Nabokov (Lycaenidae). The Karner Blue. Subspecies of the Melissa Blue of northeastern United States; protected in New York State. "A Discovery," in *Poems,* p. 15, and in *Poems and Problems,* pp. 155-56; *Pnin,* p. 128; *The Nabokov-Wilson Letters,* p. 307.

Lycaeides samuelis Nabokov. See *Lycaeides melissa samuelis* Nabokov

Lycaeides scudderi lotis Lintner = *L. argyrognomon lotis* (Lycaenidae). Subspecies of the Northern Blue which occurs only in Mendocino County, CA. *The Nabokov-Wilson Letters,* pp. 103, 126.

Lycaeides sublivens Nabokov = *L. argyrognomon sublivens* Nabokov (Lycaenidae). Subspecies of the Northern Blue occurring in the San Miguel, San Juan, and Elk Mountains, Colorado. "On a Book Called Lolita," in *Lolita,* p. 318; *Strong Opinions,* pp. 315-21 (reprint of article); *The Nabokov-Wilson Letters,* p. 265.

Lysandra cormion Nabokov (Lycaenidae). Alpes Maritimes. Nabokov places it between *Lysandra coridon* (The Chalk Hill Blue) and *Meleageria daphnis* (Meleager's Blue): "a great and delightful rarity." *Speak, Memory,* plate opposite p. 188.

/*Maculinea arion*/ (Lycaenidae). The Large Blue. This is a species that has a symbiotic relationship with ants. Central Euope, local colonies elsewhere in Europe, Russia, Asia. Extinct in England. *The Gift,* p. 122

Mann's White. *Pieris mannis* (Pieridae). Southern Europe, Morocco, Syria. *Look at the Harlequins!,* p. 35f.

Meadowbrown. *Maniola jurtina* (Satyridae). Common butterfly in Europe, North Africa, and into Asia Minor. *Laughter in the Dark,* p. 201; *Strong Opinions,* p. 60.

Monarch. *Danaus plexippus* (Danaidae). Temperate North American into South America. *Pnin,* p. 136; *Ada,* p. 158.

Morphos (Morphidae). Neotropical: Mexico to southern Brazil, northern Argentina. Except for the reference in "The Aurelian," *Details of a Sunset,* p. 105 and the mounted butterflies in *Look at the Harlequins!,* p. 67, references to this butterfly are for color or value: *Bend Sinister,* p. 227; *Pnin,* p. 92; *Ada,* p. 551.

Mulberry moths, *Bombyx mori* (Bombycidae). *Lolita,* p. 261.

Nabokovia Hemming. syn. for genus *Pseudolucia* (Lycaenidae). This "small genus" comprises one species, *faga* Dogin from Ecuador, subfamily Theclinae. Nabokov discussed it in his paper on neotropical Lycaenidae (1944). *Strong Opinions,* p. 190.

Nabokov's Pug. See *Eupithecia nabokovi* McDunnough

Nettlefly. *Aglais urticae* (Nymphalidae). European species commonly known as the Small Tortoiseshell. Larval host plant is nettle. *Look at the Harlequins!,* p. 108.

Niobe Fritillary. *Fabriciana niobe* (Nymphalidae). Continental Europe, Asia Minor. *The Gift,* p. 110.

/*Nymphalis polychloros*/ (Nymphalidae). The Large Tortoiseshell. Large portions of Europe. *Invitation to a Beheading,* p. 119.

Oak Eggars. *Lasiocampa quercus* (Lasiocampidae). Common moth of Europe to Siberia. *Speak, Memory,* p. 132.

Oleander Hawk Moth. *Daphnis nerii* (Sphingidae). Striking hawkmoth with delicate coloration and markings. Europe to India. "The Aurelian" in *Details of a Sunset,* p. 104.

Orange Moth. *Angironia prunaria* (Geometridae). Common is woods throughout much of Europe and in parts of Asia. *Speak, Memory,* p. 129.

Orange-Tip. See Zegris Butterfly.

Painted Lady. *Vanessa cardui* (Nymphalidae). Cosmopolitan. Known also as the Cosmopolite and as the Thistle Butterfly. Nabokov found it most plentiful in the Crimea and also in the Snowy Range of Wyoming. *The Gift,* pp. 123, 142; *Bend Sinister,* p. 156.

Pandora. *Pandoriana pandora* (Nymphalidae). Europe. *Look at the Harlequins!,* pp. 34-35.

Paphia Fritillary. *Argynnis paphia* (Nymphalidae). The Silver-Washed Fritillary. Europe, Asia, North Africa. *Pnin,* p. 177.

/*Papilio dardanus*/ (Papilionidae). An African swallowtail. Males of the species throughout their range have the long tails characteristic of the genus while the females vary greatly in coloration and in shape of their wings, mimicking in different localities various unpalatable species of other families of butterflies. *The Gift,* p. 123.

Pararge aegeria (Satyridae). In the text called Egerias from their species, they are known as the Speckled Wood. Most of central and northern Europe through Russia to Asia. *Speak, Memory,* p. 176.

Parnassius mnemosyne (Papilionidae). The Clouded Apollo, Pyrenees, central France, central and southern Europe to the Caucasus and central Asia. *Speak, Memory,* p. 210 and pictured on the inside covers of that book.

Parnassius phoebus golovinus (i.e., correctly *golovninus*) (Papilionidae). Subspecies of the Phoebus Parnassian or Small Apollo. High altitudes in western Alaska, Eurasia. *Speak, Memory,* p. 52.

Peacock Butterfly. *Inachia io* (Nymphalidae). Widely distributed throughout Europe, temperate Asia. *The Gift,* p. 121; *Speak, Memory,* p. 75; *Ada,* p. 524.

Peacock Moth. (1) *Saturnia pyri* (Saturnidae). Also called the Greater Emperor Moth. Europe, Asia, Africa. J. H. Fabre used this species in his experiments to prove that the female pheromone attracts males sometimes from miles away. *Invitation to a Beheading,* p. 203f; *Ada,* p. 400. (2) Probably *Antherea polyphemus* (Saturnidae), though it is not commonly known by that name. "Lines Written in Oregon," in *Poems,* p. 33 and *Poems and Problems,* p. 171.

Pear Peacock Moth. See Peacock Moth

Pearl-Bordered Fritillary. *Boloria euphrosyne* (Nymphalidae). Europe, Asia. *Speak, Memory,* p. 122.

Philotes baron (Lycaenidae). The Baton Blue. The larval food on which these small blues were found is thyme. *Despair,* p. 48.

Phytometra (Plusia) excelsa Kretschmar (Noctuidae). A "mauve-and-maroon moth. *Speak, Memory,* p. 133.

Plebejus (Lysandra) cormion Nabokov. See *Lysandra cormion* Nabokov

Poplar Admiral. See *Limenitis populi*

Purple Emperor. *Apatura iris* (Nymphalidae). Woodland species from Europe and Asia as far as Japan. *The Nabokov-Wilson Letters,* p. 113.

Puss Moth. *Cerura vinula* (Notodontidae). Palearctic "prominent" moth. *Ada,* p. 55.

Queen of Spain. *Issoria lathonia* (Nymphalidae). A fritillary known in most of Europe and parts of Asia in arid regions with scant vegetation. In *Ada,* p. 524, it is observed in Switzerland; in *Speak, Memory,* p. 218, in Russia.

Red Admirable, or Red Admiral. See *Vanessa atalanta.*

Roborovski's White. Unidentified Pieridae. *The Gift,* p. 136.

Russian Poplar Admiral. See *Limenitis populi*

Scarce Heath. *Coenonymphia hero* (Satyridae). Called the Hero Ringlet in *Speak, Memory.* Northern Europe to Asia. *Speak, Memory,* p. 132; *Strong Opinions,* p. 182.

Selene Fritillary. *Boloria selene* (Nymphalidae). Europe, Asia, North America, where it is known as the Silver-Bordered Fritillary. *The Gift,* p. 144.

Sharkmoth. *Cucullia verbasci* (Noctuidae). Also referred to as Cowl Moths in *Ada,* they are also called Monk Moth, Mullein Moth. The generic name means cowl, and the larval food plant (mullein) is also mentioned. *Ada,* p. 56.

Sievers' Carmelite. *Ptilodon (Lophopteryx) carmelita* (Notodontidae). "Just a grey moth..."

A prominent, the species is mentioned frequently in the *Entomological News* of London in the early years of this century. *Speak, Memory,* p. 132.

Silvius Skipper. *Carterocephalus silvicolus* (Hesperidae). Correctly Sylvius. The Northern Chequered Skipper. Occurs in large sections of Europe: Finland, the Baltics, northern Poland, to Germany. *Speak, Memory,* p. 132.

Small White. See Cabbage Butterfly.

Smerinthus jamaicensis (Sphingidae). Twin-Spotted Sphinx. A tentative identification on Nabokov's part from Wilson's description. Northern United States as far west as the Rocky Mountains. *The Nabokov-Wilson Letters,* p. 106.

/*Smerinthus ocellata*/ (Sphingidae). The Eyed Hawk Moth. Europe. Excellent description in the text. *Bend Sinister,* p. 134.

Spanish Orange-Tip. See Zegris Butterfly

Speckled Wood. See *Pararge aegeria*

Strymonidia W-Album (Lycaenidae). White-Letter Hairstreak. Europe. *Speak, Memory,* p. 132.

Swallowtail. *Papilio machaon* (Papilionidae). The Swallowtail of Europe. *The Gift,* p. 145; *Speak, Memory,* p. 120.

Thecla bieti =*Esakiozephrus bieti* (Lycaenidae). Tibet. *The Gift,* p. 366.

Tiger Moths. (Arctiidae). Circumpolar family of some 5000 species, many brightly colored and attractively marked. *The Gift,* p. 122.

Urania Moth. (Uraniidae). The species mentioned was probably one of the diurnal families of Africa, South America, and Australasia. Some are extremely beautiful and *Papilio* mimics. *Speak, Memory,* p. 128.

Vanessa atalanta (Nymphalidae). The Red Admiral or Red Admirable, also known as the Alderman. Holarctic and quite common. *King, Queen, Knave,* p. 44; *The Gift,* p. 36; *Pale Fire,* lines 269-71, 993; pp. 172, 290; *Speak, Memory,* p. 305; *Strong Opinions,* p. 170.

Viceroy. *Basilarchia archippus* (Nymphalidae). Throughout most of North America below the Arctic. *Ada,* p. 158.

Violet-Tinged Coppers. *Paleochrysaphanos hippothoë* (Lycaenidae). The males precede the females in emerging from winter diapause. Europe. *The Gift,* p. 145.

Virginia White. *Artogeia virginiensis* (Pieridae). The West Virginia White. Eastern United States. Called also the Toothwart White from its larval food plant *Dentaria diphylla. Pale Fire,* pp. 183f.

Yellow-Banded Ringlet. *Erebia flavofasciata* (Satyridae). Southern Swiss Alps. *Strong Opinions,* p. 136.

Yucca Moths. (Prodoxidae). Identified as *Pronuba* by Nabokov (*The Annotated Lolita,* p. 383). There are four species known to pollinate the yucca in an interdependent relationship. Humbert confused them with "whiteflies" (Hemiptera). *Lolita,* p. 158.

Zebra-Striped Swallowtail. See *Iphiclides podalirius*

Zegris Butterfly. *Zegris eupheme* (Pieridae). The Sooty Orange-Tip. Spain, also southern Russia, Iran, Asia Minor, Morocco. Other orange-tips, including those of North America, are *Anthocharis* and *Falcapica. Ada,* p. 500.

Appendix B

Butterflies and Moths: A Selected Bibliography

A. Worldwide

Dickens, Michael and Eric Storey. *The World of Moths*. New York: Macmillan, 1975.

Emmel, Thomas C. *Butterflies: Their World, Their Life Cycle, Their Behavior*. New York: Knopf, 1975.

Klots, Alexander. *The World of Butterflies and Moths*. New York: McGraw-Hill, 1958.

Lewis, H. L. *Butterflies of the World*. Chicago: Follett, 1973.

Smart, Paul. *The Illustrated Encyclopedia of the Butterfly World*. New York: Chartwell books, 1977 (Reprint of *The International Butterfly Book*, 1975).

Stanek, V. J. *Illustrated Encyclopedia of Butterflies and Moths*. London: Octopus Books, 1977.

Watson, Allan and Paul E. S. Whalley. *The Dictionary of Butterflies and Moths in Color*. New York: McGraw-Hill, 1975.

B. Europe

Brooks, Margaret and Charles Knight. *A Complete Guide to British Butterflies*. London: Jonathan Cape, 1982.

Carter, David. *Butterflies and Moths in Britain and Europe*. London: Heinemann, 1982.

Ford, E. B. *Butterflies*. Revised edition. Glasgow: Collins, 1975. (Concentration on species of England but valuable for general study.)

Higgins, L. G. and N. D. Riley. *A Field Guide to the Butterflies of Britain and Europe*. Boston: Houghton Mifflin, 1970.

Lyneborg, Leif. *Butterflies in Colour*. London: Blandford Press, 1974.

C. North America

Covell, Charles V. *Field Guide to the Moths of Eastern North America*. Boston: Houghton, Mifflin, 1984 (Peterson Field Guide Series).

Holland, William J. *The Moth Book*. Revised. New York: Dover, 1968.

Howe, William H. *The Butterflies of North America*. Garden City: Doubleday, 1975.

Klots, Alexander B. *A Field Guide to the Butterflies*. Boston: Houghton Mifflin, c1951.

Opler, Paul A. and George O. Krizek. *Butterflies East of the Great Plains*. Baltimore: Johns Hopkins University Press, 1984.

Pyle, Robert M. *The Audubon Society Field Guide to North American Butterflies*. New York: Knopf, 1981.

Bibliography

A. The Literarary Works and Letter of Vladimir Nabokov

Ada, or Ardor: A Family Chronicle. New York: McGraw-Hill, 1969.

The Annotated Lolita. Edited by Alfred Appel. New York: McGraw-Hill, 1970.

Bend Sinister. New York: McGraw-Hill, 1974.

The Defense. New York: Capricorn Books, 1970.

Despair. New York: McGraw-Hill, 1965.

Details of a Sunset, and Other Stories. New York: McGraw-Hill, 1976.

The Eye. New York: Phaedra, 1965.

The Gift, New York: Capricorn Books, 1970.

Glory. New York: McGraw-Hill, 1971.

Invitation to a Beheading. New York: Capricorn Books, 1965.

King, Queen, Knave. New York: McGraw-Hill, 1968.

Laughter in the Dark. Norfolk, Conn.: New Directions, 1960.

Lectures on Literature. Edited by Friedson Bowers. New York: Harcourt, Brace, Jovanovich, 1980.

Lolita. New York: Putnams, 1955.

Look at the Harlequins! New York: McGraw-Hill, 1974.

Mary. New York: McGraw-Hill, 1970.

___, and Edmund Wilson. *The Nabokov-Wilson Letters, 1940-1971.* Edited by Simon Karlinsky. New York: Harper, 1979.

Nabokov's Dozen. Freeport, New York: Books for Libraries, 1969. (Reprint of 1958).

Nikolai Gogol. Norfolk, Conn.: New Directions, 1944.

Pale Fire. New York: Putnams, 1962.

Pnin. Garden City, N.Y.: Doubleday, 1957.

Poems. Garden City, N.Y.: Doubleday, 1959.

Poems and Problems. New York: McGraw-Hill, 1970.

The Real Life of Sebastian Knight. Norfolk, Conn.: New Directions, 1941, 1959.

A Russian Beauty and Other Stories. New York: McGraw-Hill, 1973.

Speak, Memory, an Autobiography Revisited. Revised edition. New York: Putnams, 1966.

Strong Opinions. New York: McGraw-Hill, 1973.

Transparent Things. New York: McGraw-Hill, 1972.

A Tyrant Destroyed, and Other Stories. New York: McGraw-Hill, 1975.

Pushkin, Alexander. *Eugene Onegin.* Translated and with a Commentary by Vladimir Nabokov. 4 vols. New York: Bollingen Foundation, Pantheon Press, 1964.

B. The Lepidoptera Writings of Vladimir Nabokov

1920.   A few notes on Crimean Lepidoptera. *Entomologist* 53: 29-33.

1931. Notes on the Lepidoptera of the Pyrenées Orientales and the Ariège. *Entomologist* 64: 255-57; 268-71.

1941. On some Asiatic species of *Carterocephalus. JL N.Y. Ent. Soc.* 49: 265-67.

1942. Some new or little known Nearctic *Neonympha. Psyche* 49: 61-80.

1943. The female of *Neonympha maniola* Nabokov. *Psyche* 50: 33.

1943. The Nearctic forms of *Lycaeides* Hübner. *Psyche* 50: 87-99.

1944. Notes on the morphology of the genus *Lycaeides. Psyche* 51: 104-38.

1945. Notes on Neotropical *Plebejinae. Psyche* 52: 1-53.

1945. A third species of *Echinargus* Nabokov. *Psyche* 52: 193.

1948. A new species of *Cyclargus* Nabokov. *Entomologist* 81: 273-80.

1949. The Nearctic members of the genus *Lycaeides* Hübner. *Bull MCZ* 101: 479-641.

1952. The female of *Lycaeides sublivens* Nab. *Lepidopterists' News* 6: 35-36.

1953. Butterfly hunting in Wyoming. *Lepidopterists' News* 7: 49-52.

C. Secondary Sources

Appel, Alfred, Jr. and Charles Newman. *Nabokov, Criticisms, Reminiscences, Translations, and Tributes.* Evanston: Northwestern University Press, 1970.

Bader, Julia. *Crystal Land: Artifice in Nabokov's English Novels.* Berkeley: University of California Press, 1972.

Bruss, Elizabeth B. *Autobiographical Acts.* Baltimore: Johns Hopkins Press, 1976.

Butler, Diana. "Lolita Lepidoptera." *New World Writing,* 16 (1960), pp. 58-84.

Christensen, Inger. *The Meaning of Metaphor.* Bergen: Universitetsforlag, 1981.

Dembo, L. S., ed. *Nabokov, the Man and His Work.* Madison: University of Madison Press, 1967.

Field, Andrew. *Nabokov, His Life in Art.* Boston: Little, Brown, 1967.

____. *Nabokov, His Life in Part.* New York: Viking, 1977.

Fowler, Douglas. *Reading Nabokov.* Ithaca: Cornell University Press, 1974.

Grabes, Herbert. *Fictitious Biographies: Vladimir Nabokov's English Novels.* The Hague: Mouton, 1977.

Grayson, Jane. *Nabokov Translated.* Oxford: Oxford University Press, 1977.

Hyde, G. M. *Vladimir Nabokov, America's Russian Novelist.* London: Marion Boyars, 1977.

Lee, Lawrence L. *Vladimir Nabokov.* Boston: Twayne, 1976.

McCarthy, Mary. "A Bolt from the Blue," *New Republic,* 146 (June 4, 1962), pp. 21-27.

Mason, Bobbie Ann. *Nabokov's Garden.* Ann Arbor: Ardis, 1969.

Pilling, John. *Autobiography and Imagination: Studies in Self-Scrutiny.* London: Routledge and K. Paul, 1981.

Proffer, Carl, ed. *A Book of Things about Vladimir Nabokov.* Ann Arbor: Ardis, 1974.

____. *Keys to Lolita.* Bloomington: Indiana University Press, 1968.

Quennell, Peter, ed. *Vladimir Nabokov, a Tribute: His Life, His Work, His World.* New York: William Morrow, 1980.

Rowe, William W. *Nabokov's Deceptive World.* New York: New York University Press, 1971.

____. *Nabokov's Spectral Dimension.* Ann Arbor: Ardis, 1981.

Stegner, Page. *Escape into Aesthetics: The Art of Vladimir Nabokov.* New York: Dial Press, 1966.

Stuart, Dabney. *Nabokov: Dimensions of Parody.* Baton Rouge: Louisiana State University Press, 1978.